Accelerating Angular Development with Ivy

A practical guide to building faster and more testable Angular apps with the new Ivy engine

Lars Gyrup Brink Nielsen

Jacob Andresen

BIRMINGHAM—MUMBAI

Accelerating Angular Development with Ivy

Associate Group Product Manager: Pavan Ramchandani
Publishing Product Manager: Pavan Ramchandani
Senior Editor: Sofi Rogers
Content Development Editor: Rakhi Patel
Technical Editor: Shubham Sharma
Copy Editor: Safis Editing
Project Coordinator: Manthan Patel
Proofreader: Safis Editing
Indexer: Subalakshmi Govindhan
Production Designer: Roshan Kawale

First published: September 2021

Production reference: 1300921

Published by Packt Publishing Ltd.
Livery Place
35 Livery Street
Birmingham
B3 2PB, UK.

ISBN 978-1-80020-521-5

www.packt.com

This book is dedicated to my daughters, for inspiring me to always do better. My greatest thanks to my wife, for supporting me and my extra-curricular activities such as writing this book.

– Lars Gyrup Brink Nielsen

I dedicate this book to the Angular community.

– Jacob Andresen

Foreword

Angular Ivy was first introduced in Angular version 9 in 2020 and it has been kind of a black box for many developers. Few developers know much about Ivy but *Lars Gyrup Brink Nielsen* is one of the leading Ivy experts in the world. This is the first book that goes into more detail about Ivy features and tooling. It is a must-read for all developers at all levels. Even if you are an expert in Angular, you will learn a lot from this book.

Santosh Yadav

Google Developer Expert in Angular, GitHub Star, Auth0 Ambassador,
Cofounder of This is Angular and Software Consultant.

Contributors

About the authors

Lars Gyrup Brink Nielsen is a frontend architect at Systemate A/S in Denmark. As a cofounder of the open learning non-profit This is Learning, he creates open platforms for free knowledge and public learning. Lars is a tech writer, international tech speaker, FOSS maintainer, Microsoft MVP in Developer Technologies, and GitHub Star.

I want to extend my thanks to my computer science professor, Hans Hüttel, for being an inspiration early in my career and for teaching me regular expressions.

Jacob Andresen works as a senior software developer based in Copenhagen, Denmark. He has been working as a software developer and consultant in information retrieval systems and web applications since 2002.

I want to thank my wife, Anita, and my daughter, Sofie, for understanding why I spend all those hours in front of a computer screen.

About the reviewer

Anu Nagan G has worked in various corporate organizations, from a SaaS start-up (**GenDeep**), to midsize (**GAVS**), and Fortune 500 companies (**DXC**) playing various roles there, such as technical product manager, full stack product lead (Angular, Java, Python, and AWS], and delivery lead, respectively, in his 8-year tenure. Currently, he is with Lister Technologies leading parallel projects such as advanced AI and analytics product CortexAI, clinical mobile app development, and Salesforce automation for B2B businesses. He has contributed to various AIOps products, including ZIF and Gcare in the past. He is an avid reader and cinephile who loves to play guitar and makes short films with his friends.

I would like to thank my wife, Hema, for always giving me the freedom to pursue my interests.

Table of Contents

3

Introducing CSS Custom Properties and New Provider Scopes

4

Exploring Angular Components Features

5

Using CSS Custom Properties

6

Using Angular Components

7

Component Harnesses

8

Additional Provider Scopes

9

Debugging with the New Ivy Runtime APIs

10

Using the Angular Compatibility Compiler

11

Migrating Your Angular Application from View Engine to Ivy

12

Embracing Ahead-of-Time Compilation

Other Books You May Enjoy

Index

Preface

Angular Ivy is the current generation of Google's open source framework. Angular is well known for being bundled with a set of tools to increase application robustness and boost the developer experience. Angular relies on and evolves with TypeScript, the open source typed JavaScript superset by Microsoft.

Accelerating Angular Development with Ivy is the result of thousands of hours of research and experiments to unveil the most exciting runtime APIs, testing APIs, debugging APIs, and tooling that are introduced by Angular Ivy. The authors have carefully selected topics that are widely applicable to different types of Angular applications. Most of the topics discussed in this book have in common that they are largely uncovered by official documentation or other publicly available content.

This book contains three parts. *Part 1, A Quick and Functional Guide to Angular Ivy*, approaches new features, testing APIs and tooling from a theoretical and informative perspective. Significant topics are put in a business or application context through simple or extended examples to help you discover their value and inspire future use cases.

Part 2, Build a Real-World Application with the Angular Ivy Features You Have Learned, gets your hands dirty with some of the most exciting topics that were introduced in *Part 1*. We use the real-world application *Angular Academy* as a starting point and extend it with features through practical step-by-step exercises.

Part 3, Upgrade Your View Engine Application and Development Workflow to Angular Ivy, is an extensive migration guide for updating your application from Angular View Engine to Angular Ivy. As a hidden gem, you will learn about debugging your application at runtime by using Angular Ivy's debugging API on a component. Additionally, *Part 3* introduces the Angular Compatibility Compiler and teaches you about optimizations. *Part 3* is also the part to refer to to learn about the effect of the Angular Ahead-of-Time compiler and how to work around its edge cases and limitations.

After reading this book, you will have an overview of significant Angular Ivy features. Through realistic examples, you will have gained unique insights into fresh APIs and use cases. You will know the impact Angular Ivy has on your application and how you can maximize its benefits. After discovering how Angular Ivy improves your developer experience through all phases of developing your application, you will have boosted your productivity, and this book will serve as an excellent reference for every Angular Ivy application you develop in the future.

Who this book is for

Accelerating Angular Development with Ivy is written for the experienced Angular developer who wants to catch up on the exciting unique features and tooling added or improved by Angular Ivy. This book covers Angular versions 9 through 12.

What this book covers

Chapter 1, Discovering New APIs and Language Syntax, first dives into three important additions to TypeScript in versions supported by Angular Ivy. The optional chaining and nullish coalescing operators are discussed and shown in a business context separately. Perhaps more importantly, their powerful combination is demonstrated.

This chapter goes on to optimize regional support through Angular Ivy's improved globalization APIs, briefly discussing how to configure the Angular CLI to bundle locale data and translation files. Together, we implement a bi-directional media directive for art direction and combine it with a locale picker that lazily loads any locale supported by Angular. This is a testament to the power of combining Angular with browser APIs.

Testability is an important concern for Angular. This chapter is where we uncover how Angular's `TestBed.inject` method adds stronger typing to tests by exploring its API and a simple example. Additionally, we use an example to learn about Angular Material's `FakeMatIconRegistry` stub service and how it is provided in an Angular testing module.

Chapter 2, Boosting Developer Productivity Through Tooling, Configuration, and Convenience, discusses Angular Ivy's style binding precedence rules and how they make style evaluation predictable. An elaborate example lets us explore these rules.

This chapter discusses which metadata can be shared through directive and component inheritance. This is covered through examples that demonstrate realistic use cases.

The Angular Ahead-of-Time compiler and the Angular Compatibility Compiler are introduced as an integral part of Angular Ivy, affecting all phases of development.

This chapter also discusses Angular Ivy's Strict mode and related compiler options. We also cover TypeScript's error-expecting compiler annotations, which are useful for unit tests.

Finally, Angular Ivy's improved compilation errors are demonstrated and compared to those of Angular View Engine. Additionally, we discuss tooltips added to the Angular Language Service by Ivy.

Chapter 3, Introducing CSS Custom Properties and New Provider Scopes, introduces CSS Custom Properties and their scoping before demonstrating novel techniques for interacting with CSS Custom Properties through Angular data binding.

Angular Ivy introduces the any and platform provider scopes. Both new provider scopes have limited but important and advanced use cases. Each of them is discussed in detail and example dependency hierarchies are visualized to facilitate understanding. To understand an important use case of the any provider scope, a web API configuration example is covered step by step. The platform provider scope is useful for Angular Elements web components and Angular microfrontends. A partial microfrontend is introduced to demonstrate how dependency injection works when using the platform provider scope.

Chapter 4, Exploring Angular Components Features, covers novel APIs and packages introduced by Ivy to Angular's UI packages owned by the Angular Components team. Through extensive API references and basic examples, you will learn about the Angular YouTube Player and the most common parts of the Angular Google Maps components package. The Clipboard API is added to Angular CDK by Ivy. We discuss the directive, service, and class it contains.

Finally, Angular's new test-as-a-user API, known as *component harnesses*, is explored. The basic concepts *harness environments*, *harness loaders*, and *component harnesses* are introduced. We walk through the Angular Material Button component harness API before using it in practice for an example component that also relies on the Angular Material Select component harness.

Chapter 5, Using CSS Custom Properties, introduces the Angular Academy app and shows how to create a theme component that can control the look and feel of the app using CSS Custom Properties. Furthermore, the chapter discusses how to use CSS Custom Properties for a flexible grid layout.

Chapter 6, Using Angular Components, walks through how the Angular Academy app is structured and shows how to implement and use the new Angular components in a modular fashion.

Chapter 7, Component Harnesses, further dives into the details of the Angular testing harnesses and how to apply the Material testing harnesses in the context of the Angular Academy app. A custom Video Test harness is introduced to illustrate how to approach layered testing in the context of the Angular Academy app.

Chapter 8, Additional Provider Scopes, illustrates how to use the any provider scope to implement a configurable ThemeService in the context of the Angular Academy app. Additionally, we show how to share information across applications using the platform provider scope.

Chapter 9, Debugging with the New Ivy Runtime APIs, first describes Angular Ivy's debugging API in full. However, it is the most widely applicable debugging utility functions, `ng.applyChanges`, `ng.getComponent`, `ng.getContext`, and `ng.getListeners`, that are covered in detail by debugging a component and its related directive. This will teach you how to inspect an active component as well as its event listeners and embedded view context. These are important techniques to master when developing Angular applications.

Chapter 10, Using the Angular Compatibility Compiler, introduces the Angular Compatibility Compiler and discusses its role as a temporary part of Angular Ivy throughout the Angular View Engine to Ivy transition period. Significant parameters for its CLI are described before recommendations are made to optimize the compiler for local development.

This chapter further discusses Angular Compatibility Compiler optimization techniques for CI/CD workflows and monorepo workspaces as a result of extensive research, experiments, and measurements.

Chapter 11, Migrating Your Angular Application from View Engine to Ivy, is a unique and extensive migration guide for Angular Ivy. First, we discuss the phases and tools used in the Angular update process and how to manage Angular dependencies. Next, we discuss how to perform automated Angular Ivy migrations and how to review the most significant during or after Ivy migration. Additionally, an optional but recommended migration is described.

This chapter recommends manual Ivy migrations to future-proof your Angular application. These migrations are related to navigation, change detection, and unit tests.

Chapter 12, Embracing Ahead-of-Time Compilation, focuses on the Ahead-of-Time Angular compiler. We discuss its impact on all phases of developing your Angular application before introducing techniques for dealing with concrete limitations and edge cases of Ahead-of-Time compilation.

This chapter contains a unique section that introduces two techniques for initializing asynchronous dependencies. We use feature flags as a case study for this.

To get the most out of this book

This book assumes experience with Angular version 8 or earlier. The book is also applicable to developers working on the latest Angular version. You should be familiar with the Angular CLI to start the applications that are in this book's companion GitHub repository.

To focus on exciting additions and improvement, we skip all introductions to core concepts and building blocks of Angular. Therefore, we assume that you feel comfortable developing Angular directives, components, and services.

Angular's dependency injection system is integral to any Angular application. We expect you to be familiar with the APIs for managing dependencies.

Software covered in the book	Operating system requirements
Angular versions 9 through 12	Windows, macOS, or Linux
TypeScript versions 3.6 through 4.3	Windows, macOS, or Linux
Angular CLI	Windows, macOS, or Linux
Angular CDK	Windows, macOS, or Linux
Angular Google Maps	Windows, macOS, or Linux
Angular Material	Windows, macOS, or Linux
Angular YouTube Player	Windows, macOS, or Linux

Make sure to install a recent version of Angular CLI globally.

If you are using the digital version of this book, we advise you to type the code yourself or access the code from the book's GitHub repository (a link is available in the next section). Doing so will help you avoid any potential errors related to the copying and pasting of code.

Download the example code files

You can download the example code files for this book from GitHub at `https://github.com/PacktPublishing/Accelerating-Angular-Development-with-Ivy`. If there's an update to the code, it will be updated in the GitHub repository.

We also have other code bundles from our rich catalog of books and videos available at `https://github.com/PacktPublishing/`. Check them out!

Download the color images

We also provide a PDF file that has color images of the screenshots and diagrams used in this book. You can download it here: `https://static.packt-cdn.com/downloads/9781800205215_ColorImages.pdf`.

Conventions used

There are a number of text conventions used throughout this book.

`Code in text`: Indicates code words in text, database table names, folder names, filenames, file extensions, pathnames, dummy URLs, user input, and Twitter handles. Here is an example: "Mount the downloaded `WebStorm-10*.dmg` disk image file as another disk in your system."

A block of code is set as follows:

```
ngOnInit(): void {
    this.document.body.appendChild(
     this.#youtubeIframeScript);
  }
```

Any command-line input or output is written as follows:

```
ng add @angular/youtube-player
```

Bold: Indicates a new term, an important word, or words that you see onscreen. For instance, words in menus or dialog boxes appear in **bold**. Here is an example: "The user can do this by selecting a color by going to the **Color Input** field to obtain the desired color for the given **Header background** setting."

> **Tips or important notes**
> Appear like this.

Get in touch

Feedback from our readers is always welcome.

General feedback: If you have questions about any aspect of this book, email us at `customercare@packtpub.com` and mention the book title in the subject of your message.

Errata: Although we have taken every care to ensure the accuracy of our content, mistakes do happen. If you have found a mistake in this book, we would be grateful if you would report this to us. Please visit `www.packtpub.com/support/errata` and fill in the form.

Piracy: If you come across any illegal copies of our works in any form on the internet, we would be grateful if you would provide us with the location address or website name. Please contact us at `copyright@packt.com` with a link to the material.

If you are interested in becoming an author: If there is a topic that you have expertise in and you are interested in either writing or contributing to a book, please visit `authors.packtpub.com`.

Share Your Thoughts

Once you've read *Accelerating Angular Development with Ivy*, we'd love to hear your thoughts! Scan the QR code below to go straight to the Amazon review page for this book and share your feedback.

https://packt.link/r/180020521X

Your review is important to us and the tech community and will help us make sure we're delivering excellent quality content.

1
Discovering New APIs and Language Syntax

As its only officially supported programming language, Angular is tightly coupled with TypeScript. Support for new versions of TypeScript is introduced with major and minor version releases of Angular. In this chapter, we will explore three powerful language features that have been released in the recent versions of TypeScript and Angular:

- The optional chaining operator (`?.`)
- The nullish coalescing operator (`??`)
- Native private class members (`#`)

Through simple examples, we will highlight the strengths of these modern programming language features. We will even learn how two of the new operators work elegantly together in common scenarios. Learning about this new syntax and semantics is key to understanding the examples throughout this book.

Globalization is the process of supporting and adapting multilingual and regional capabilities in an application. Angular Ivy introduces improved globalization APIs. Together, we will learn about localization bundling, directionality querying, and lazy loading locale data through comprehensive examples.

Many core parts of Angular are built with testability in mind. Angular Ivy introduces strongly typed dependency resolving in tests and a fake icon registry for integrated component tests.

In this first chapter, we will cover the following topics:

- Modernizing your code with powerful language features
- Optimizing regional support with improved globalization APIs
- Enhancing tests with stronger types and new APIs

The demo application featured in *Part 2, Build a Real-World Application with the Angular Ivy Features You Learned* does not use globalization APIs. The tests featured in that part focus on component testing using the new concept component testing harnesses, which will be introduced in *Chapter 3, Introducing CSS Custom Properties and New Provider Scopes.*

After reading this chapter, you will be able to refactor your existing Angular applications and tests to use modern standards and Angular APIs when developing.

Technical requirements

To support all the features introduced in this chapter, your application requires at least the following:

- Angular Ivy version 9.1
- TypeScript version 3.8

You can find the complete code examples for the globalization APIs in this book's companion GitHub repository at `https://github.com/PacktPublishing/Accelerating-Angular-Development-with-Ivy/tree/main/projects/chapter1/globalization`.

Modernizing your code with powerful language features

TypeScript is an integral part of Angular, but because Angular has its own compiler transformations that extend TypeScript's compilation steps, we are inevitably tied to the version of TypeScript that the Angular compiler supports. Fortunately, Angular is good at keeping up with recent versions of TypeScript. In this section, we are going to discuss some of the most noteworthy additions to the TypeScript language in the most recent versions supported by Angular Ivy.

The optional chaining operator

TypeScript version 3.7 introduces a new operator for optional property access, optional element access, and optional calls. The optional chaining operator (`?.`) short circuits in the case of nullish values – that is, `null` or `undefined` – in which case it evaluates to `undefined`.

Optional chaining is great for working with composite objects or just plain old complex data structures such as large business documents transferred from a server, a dynamic runtime configuration, or telemetry from **Internet of Things** (**IoT**) devices.

The optional chaining operator allows us to be more concise in a single expression that attempts to access a hierarchy of properties that may or may not be available.

Say we are working on a document processing system that supports **Universal Business Language** (**UBL**) documents in JSON format. In our UBL invoice parser, we want to determine the UBL version that the document follows to be able to parse it according to a standard version. We fall back to UBL version 2.1 if it's left unspecified.

Without optional chaining, our code might look like this, given an `invoice` variable containing the invoice document:

```
const ublVersion =
  (invoice.Invoice[0].UBLVersionID &&
    invoice.Invoice[0].UBLVersionID[0] &&
    invoice.Invoice[0].UBLVersionID[0].IdentifierContent) ||
    '2.1';
```

For each optional property or array element in the invoice data structure, we must make a conditional check to break the circuit; otherwise, we will end up trying to access a property of an `undefined` value, which will result in a runtime error.

With the optional chaining operator, our statement is simplified to the following one:

```
const ublVersion =
  invoice.Invoice[0].UBLVersionID?.[0]?.IdentifierContent ||
  '2.1';
```

Now, we can see the structure of the document and identify every optional property or array element as they are succeeded by the optional chaining operator, `?.`.

The noise of conditional access is reduced to a bare minimum. At the same time, the risk of a mistake occurring is reduced as it is less code overall and there is no repeated code, meaning that we would only have to make a single change in our code if the shape of the data structure changed.

In the invoice document examples in the previous two code listings, we have seen the optional chaining operator being used for optional property and element access.

> **Important Note**
> Optional access using strings and symbols is also supported. Even optional access using computed property names is supported.

The last use case for this new operator is optional function or method calls. Let's go through an example.

Let's say we follow a classical API style of allowing our consumers to pass success and error callbacks. If the error callback is optional, we must check that it was passed before calling it; otherwise, we will get a runtime error.

In the `catch` clause of the following code, we can see a conditional call for the `onError` callback:

```
function parseDocument(
  json: string,
  onSuccess: (document: UBLDocument) => void,
  onError?: (error: Error) => void
): void {
  try {
    const document = JSON.parse(json);
    onSuccess(document);
  } catch (error) {
    if (onError) {
```

```
      onError(error);
    }
  }
}
```

With the optional chaining operator, we can simplify the conditional function call as seen in the following code:

```
function parseDocument(
  json: string,
  onSuccess: (document: UBLDocument) => void,
  onError?: (error: Error) => void
): void {
  try {
    const document = JSON.parse(json);
    onSuccess(document);
  } catch (error) {
    onError?.(error);
  }
}
```

Even though a function is not a member of an object, we can still use an optional call to invoke it conditionally. We put the optional chaining operator between the function name and the parentheses that wrap the arguments passed to it, as seen in the previous code.

Keep in mind that the optional chaining operator returns `undefined` when the conditional property it guards contains a nullish value. This means that an expression that involves optional chaining can only return `null` if the final property, element, or return value contains `null`.

Another caveat is that using optional chaining can lead to unexpected results when short-circuited, as demonstrated next.

In the following code, we are accessing the `temperature` property of the object in the `telemetry` variable. If the temperature property is non-nullish, we can access its `celsius` property to read the degrees in Celsius, which gives us the answer we expect when we use a formula to convert it into degrees in Fahrenheit:

```
const degreesFahrenheit = telemetry.temperature?.celsius * 1.8
+ 32;
```

However, when the optional property access meets a nullish value in the `temperature` property, the first expression on the right-hand side of the expression will evaluate to `undefined`. The full calculation evaluates to `NaN` as it is evaluated as `undefined * 1.8 + 32`.

The nullish coalescing operator

The new nullish coalescing operator (`??`) was introduced in TypeScript version 3.7. When its left-hand side expression evaluates to a nullish value – that is, `null` or `undefined` – it will evaluate the full expression to its right-hand side expression. This is often used to declare default values.

We already know the default parameter syntax for functions. A nullish coalescing operation is similar but can be used in expressions. Unlike default parameters, nullish coalescing evaluates to its right-hand side when its left-hand side is evaluated to `null`.

Let's look at an example of declaring a default option without nullish coalescing:

```
function changeSettings(settings: {
  sleepTimer: number;
  volume: number;
}): void {
  const volumeSetting = settings.volume || 0.5;
  sendVolumeSignal(volumeSetting);

  const sleepTimerSetting = settings.sleepTimer || 900;
  sendSleepTimerSignal(sleepTimerSetting);
}
```

As we can see, we used the logical OR operator (`||`) to declare a default value. The problem with this JavaScript idiom is that it ignores *all* falsy values. This includes `null` and `undefined`, but also zero (`0`), empty string (`''`), `false`, `NaN`, and big integer zero (`0n`).

As you might have realized, this introduces a subtle bug in the `changeSettings` function from the previous code block. If we pass zero as the volume setting to mute the device, the logical OR operation will default to `0.5`, effectively preventing anyone from ever turning the volume off entirely.

We are going to rewrite this function with the nullish coalescing operator to fix this bug, as shown here:

```
function changeSettings(settings: {
  sleepTimer: number;
  volume: number;
}): void {
  const volumeSetting = settings.volume ?? 0.5;
  sendVolumeSignal(volumeSetting);

  const sleepTimerSetting = settings.sleepTimer ?? 900;
  sendSleepTimerSignal(sleepTimerSetting);
}
```

Now, the function's consumers can mute the device as passing zero as the volume setting will not evaluate to the default value of `0.5`.

Handling nullish values with the new operators

The `null` and `undefined` nullish values are generic bottom values of the JavaScript language. As an example, the JSON format does not include `undefined`. Instead, any property with an undefined value is left out to save bandwidth when transmitted and space when stored. However, many server-side programming languages have a null type that converts this bottom value into JavaScript's `null` value.

Some of the APIs of JavaScript use the nullish values interchangeably, while others only accept one of them or differ in behavior if one is passed compared to the other. This leaves room for a lot of confusion, mistakes, and bugs.

TypeScript is potentially a strongly typed language but its typing can be relaxed, and it is up to us, as developers, to identify where values of unexpected types can enter our system.

For a long time, we have used the idiom of comparing a value to `null` using a loose equality check, as shown in the following code, to handle both nullish values:

```
function parseNumber(value?: string | null): number {
  return value == null
    ? NaN
    : parseFloat(value);
}
```

With the optional chaining and nullish coalescing operators, we can add another pair of tools to our toolbox.

Let's say we have a function that takes an `options` object and we want to read an option or fall back to a default value, as shown in the following example:

```
function prettyPrint<T extends {}>(
  value: T,
  options?: { spaces?: number }
): void {
  const spaces = options?.spaces ?? 2;
  const json = JSON.stringify(value, undefined, spaces);

  console.log(json);
}
```

As we can see, we can combine optional chaining with nullish coalescing. Had we not added a nullish coalescing operation, `undefined` might have been assigned to the `spaces` variable.

Had the consumer passed `0` (zero) as the `spaces` option while we used the logical OR operator (`||`) instead of nullish coalescing, the provided option would be overridden by our default value, `2`.

> **Important Note**
>
> Both of the nullish operators are part of the ECMAScript standard, which ratifies the JavaScript language. They are not unique to TypeScript.

Native private class members

We are already familiar with TypeScript's access modifiers; that is, `private`, `protected`, and `public`. TypeScript version 3.8 introduced private class members, which can't be accessed from outside the class scope they are defined in, even at runtime. This is different from the `private` access modifier, which is only evaluated at compile time.

Private class fields are a feature that TypeScript bases on an ECMAScript proposal that has not been included in the ECMAScript standard, as of ECMAScript 2021:

```
class Person {
  #name: string;

  get name() {
    return this.#name;
  }

  constructor(name: string) {
    this.#name = name;
  }
}
```

In the preceding code, we can see the `Person` class, which has a private `#name` field that we expose as read-only through the public `name` getter. The only way to set the name value is through the class constructor, which initializes the private `#name` field.

The `Employee` class in the following code extends `Person`, but can only set the name by forwarding a `string` value to the `Person` constructor by calling `super(name)`:

```
class Employee extends Person {
  #salary: number;

  constructor(name: string, salary: number = 0) {
    super(name);
    this.#salary = salary;
  }

  async giveRaise(raise: number): Promise<void> {
    this.#salary += raise;
    await this.#storySalary();
  }

  pay(): void {
    console.log(`¤ ${this.#salary} is paid to ${this.name}.`);
  }
```

```
  async #storeSalary(): Promise<void> {
    console.log(`Salary ¤ ${this.#salary} is stored for ${this.
    name}`);
  }
}
```

In the preceding code, we can see how the `Employee` superclass can access the read-only `name` property it inherits from the `Person` class. This is seen both in the public `pay` method and the private `#storeSalary` method.

If we were to instantiate an employee from the `Employee` class in the previous code listing, we would have no way of accessing the `#name` and `#salary` private fields. We are encapsulating the data in private fields and only allowing operations through public methods and properties.

If we try to access the private field from outside the employee instance at runtime, we will get an error. This is not the case if we use the `private` access modifier. However, TypeScript catches both access errors during compilation.

Private, protected, and public properties can be declared and initialized in a class constructor, but private fields cannot. It is also not possible to leave out the private field declarations. Their names must be declared, even if we do not initialize them during construction.

It is possible to declare and use private members in the case of static and instance fields, methods, getters, and setters. With private members, we can shadow parent members in subclasses. Both classes will encapsulate their separately accessible and assignable values that share the same name. This is not possible with TypeScript access modifiers.

To use private members with TypeScript, we must target at least ECMAScript 2015 in our output as their transpiled output uses `WeakMap`.

Now that we are familiar with the use cases and caveats of the optional chaining and nullish coalescing operators, as well as the difference between TypeScript access modifiers and native private class members, let's explore the changes to Angular's globalization APIs and tooling.

Optimizing regional support with improved globalization APIs

Multilingual applications use globalization to give users from different countries and backgrounds a regional experience. Angular has built-in APIs for managing both internationalization and localization. In this section, we will walk through configuration and implementation examples to illustrate some of the new globalization possibilities Ivy brings us.

Bundling localizations

Angular uses locale data for regional variations for formatting dates, currencies, decimal numbers, and percentages. In Angular Ivy, we can bundle a single locale as part of the build by using the `localize` builder option. Let's say we wanted our application to use the French locale. We can do this by configuring our project build as follows. This also implicitly replaces the value provided for the `LOCALE_ID` dependency injection token in the `@angular/core` package:

```
{
  "projects": {
    "my-app": {
      "projectType": "application",
      "i18n": {
        "locales": {
          "fr": "src/locales/translations.fr.xlf"
        }
      },
      "architect": {
        "build": {
          "builder": "@angular-devkit/build-angular:browser",
          "options": {
            "localize": ["fr"]
          }
        }
      }
    }
  }
}
```

The `localize` option identifies that the French locale (`"fr"`) is bundled and loaded with the application and that the `locales.fr` option specifies the path of the French translation file.

When migrating to Ivy, we also need to run `ng add @angular/localize` to enable Angular's built-in globalization APIs. If we don't use those, we can leave this package out to prevent increasing the bundle's size.

We can also choose to create a build configuration per locale, but we will leave that as an exercise for you. A build configuration per locale is a good choice if we are using Angular's built-in globalization APIs. The localization build speed has improved significantly in Ivy.

Lazy loading locale data

Before Angular Ivy, we had to load and register data for all the locales we wanted to support in our application, except for the `"en-US"` locale, which was always bundled. We did that using the `registerLocaleData` function.

When using an application build that comes with a bundled locale and translations, `registerLocaleData` is still the way to go. However, if we are not using Angular's built-in internationalization APIs, we can choose to lazy load locale data, which sits well with dynamic translation libraries such as ngx-translate and Transloco.

To lazy load locale data, we can dynamically import the `@angular/common/locales/global/fr` subpackage for French locale data. The following code shows a class-based service that we can use with a locale picker:

```
import { Injectable } from '@angular/core';
import { EMPTY, from, Observable } from 'rxjs';
import { catchError, mapTo } from 'rxjs/operators';

import { LanguageTag } from '../../shared/ui/language-tag';

@Injectable({
  providedIn: 'root',
})
export class LocaleLoader {
  load(locale: LanguageTag): Observable<void> {
    return from(import(`@angular/common/locales/
    global/${locale}`)).pipe(
      catchError(() => {
```

```
        console.error(`Error when loading locale "${locale}"`);

        return EMPTY;
      }),
      mapTo(undefined)
    );
  }
}
```

The `LanguageTag` type represents a text string that follows the **BCP47** format (`https://www.ietf.org/rfc/bcp/bcp47.txt`). The `LocaleLoader` service uses the dynamic function-like `import` statement to lazy load the locale data and report any errors to the browser console. This is all wrapped in an observable.

Let's walk through the locale picker template:

```
<label>
  Pick a locale
  <input list="locales" [formControl]="selectedLocale" />
</label>
<datalist id="locales">
  <option *ngFor="let locale of locales" [value]="locale"> </
  option>
</datalist>
```

Its template uses the native `<datalist>` element to enable autocomplete in its text box. The `selectedLocale` form control is bound to the `<input>` element and two validators are added:

```
import {
  ChangeDetectionStrategy,
  Component,
  Input,
  Output,
} from '@angular/core';
import { FormControl, Validators } from '@angular/forms';
import { Observable } from 'rxjs';

import { LanguageTag } from '../../shared/ui/language-tag';
```

```
import { validValueChanges } from '../../shared/ui/valid-value-changes';
import { localeValidator } from './locale-validator';

@Component({
  changeDetection: ChangeDetectionStrategy.OnPush,
  selector: 'app-locale-picker',
  styles: [':host { display: block; }'],
  templateUrl: './locale-picker.component.html',
})
export class LocalePickerComponent {
  @Input()
  set locale(value: LanguageTag) {
    this.selectedLocale.setValue(value, {
      emitEvent: false,
      emitViewToModelChange: false,
    });
  }
  @Input()
  locales: ReadonlyArray<LanguageTag> = [];
  @Output()
  localeChange: Observable<LanguageTag> = validValueChanges(
    this.selectedLocale
  );
  selectedLocale = new FormControl('', [Validators.required,
  localeValidator]);

}
```

`localeValidator` in the previous code block is composed of two validators, as shown in the following code:

```
import {
  AbstractControl,
  ValidationErrors,
  ValidatorFn,
  Validators,
```

```
} from '@angular/forms';

import { languageTagPattern } from '../../shared/ui/language-
tag';
import { allLocales } from './all-locales';

const knownLocaleValidator: ValidatorFn = (
  control: AbstractControl
): ValidationErrors | null => {
  const isValid = allLocales.includes(control.value);

  return isValid ? null : { locale: true };
};
export const localeValidator = Validators.compose([
  Validators.pattern(languageTagPattern),
  knownLocaleValidator,
]) as ValidatorFn;
```

This validator verifies that the entered text string matches the language tag format, as specified by BCP47 (`https://www.ietf.org/rfc/bcp/bcp47.txt`), by using the regular expression listed in the following code:

```
export const languageTagPattern = /^[a-z]{2,3}(-[A-Z]{1}[a-z]
{3})?(-[A-Z]{2}|\d{3})?$/;
```

Finally, `localeValidator` looks up the language tag in a list of known locales.

`LocalePickerComponent` is a presentational component – it does not directly trigger any side effects when the user picks a locale. Instead, it emits the selected locale through its output property, `localeChange`. This is done by passing its form control to the `validValueChanges` utility function, which is listed in the following code block:

```
import { FormControl } from '@angular/forms';
import { Observable } from 'rxjs';
import { filter, map, withLatestFrom } from 'rxjs/operators';

import { ValidationStatus } from './validation-status';

export function validValueChanges<T>(control: FormControl):
Observable<T> {
```

```
  return (control.statusChanges as
  Observable<ValidationStatus>).pipe(
    filter((status) => status === 'VALID'),
    withLatestFrom(control.valueChanges as Observable<T>),
    map(([_, value]) => value)
  );
}
```

When passed a form control, `validValueChanges` composes a stream of valid values that are emitted by that form control.

The locale picker is shown in the following screenshot:

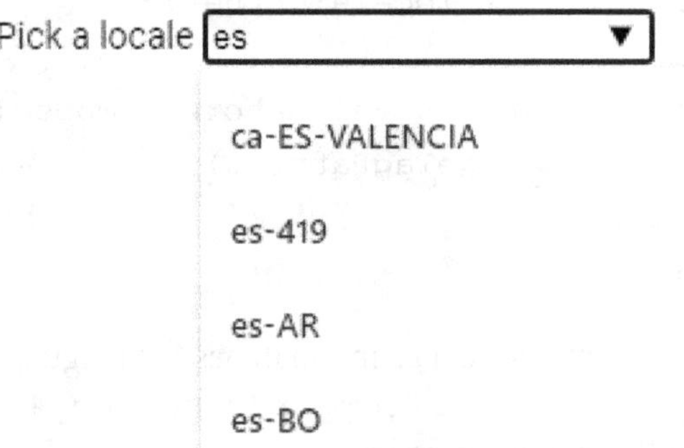

Figure 1.1 – Locale picker, as shown in Google Chrome

As you might have guessed, we need a container component to trigger the locale loader service. A reference solution for this can be found in this book's companion GitHub repository, which was referenced in the introduction to this chapter.

Dynamically setting the dir attribute based on locale direction

Ivy introduces the `getLocaleDirection` function in the `@angular/common` package to query the direction of any available locale data. Let's use this to dynamically set the `dir` attribute on our root component's host element.

Because we cannot apply a directive to the root component, we must create a service that we can provide and inject into the root component to run its initial side effects, as shown in the following code:

```
import { Component } from '@angular/core';

import { HostDirectionService } from './shared/ui/host-
direction.service';

@Component({
  selector: 'app-root',
  template: '<app-locale></app-locale>',
  viewProviders: [HostDirectionService],
})
export class AppComponent {
  constructor(
    // Inject to eagerly instantiate
    hostDirection: HostDirectionService
  ) {}
}
```

Given that we have a locale state service that exposes an observable `locale$` property, we can create the host direction service listed in the following code:

```
import { Direction } from '@angular/cdk/bidi';
import { getLocaleDirection } from '@angular/common';
import { ElementRef, Injectable, OnDestroy, Renderer2 } from '@
angular/core';
import { Observable, Subject } from 'rxjs';
import { map, takeUntil } from 'rxjs/operators';

@Injectable()
export class HostDirectionService implements OnDestroy {
  #destroy = new Subject<void>();
  #direction$: Observable<Direction> = this.localeState.
  locale$.pipe(
```

```
    map((locale) => getLocaleDirection(locale))
  );

  constructor(
    private localeState: LocaleStateService,
    private host: ElementRef<HTMLElement>,
    private renderer: Renderer2
  ) {
    this.#direction$
      .pipe(takeUntil(this.#destroy))
      .subscribe((direction) => this.
      setHostDirection(direction));
  }

  ngOnDestroy(): void {
    this.#destroy.next();
    this.#destroy.complete();
  }

  private setHostDirection(direction: Direction): void {
    this.renderer.setAttribute(this.host.nativeElement, 'dir',
    direction);
  }
}
```

This service uses the `Renderer2` service to set the value of the `dir` attribute of the host element. The host element is accessed through constructor injection.

We could easily create a similar service to set the value of the `lang` attribute on the host element. This is left as an exercise for you, but you can find a reference solution in this book's companion GitHub repository.

Art direction using the directionality query

You might know that `<picture>` elements can be combined with `<source>` elements that contain media queries to lazy load responsive images based on, for example, viewport sizes.

Would it not be cool if we could lazy load images based on the direction of the current locale? The following code shows what that could look like in an Angular application:

```
<picture>
  <source
    *media="'(dir: ltr)'"
    srcset="https://via.placeholder.com/150?text=LTR"
    />
  <source
    *media="'(dir: rtl)'"
    srcset="https://via.placeholder.com/150?text=RTL"
    />
  <img src="https://via.placeholder.com/150?text=LTR" />
</picture>
```

Here, we can create a media query-like syntax that we can pass to the `media` attribute, which is a structural directive that conditionally inserts and removes this element based on the current locale's direction, as we will see in the following code blocks:

1. First, we have the `import` statements, including `LocaleStateService`, which we looked at in the previous sections. The `directionQueryPattern` regular expression is used to match the `Direction` query:

```
import { Direction } from '@angular/cdk/bidi';
import { getLocaleDirection } from '@angular/common';
import {
  Directive,
  EmbeddedViewRef,
  Input,
  OnDestroy,
  OnInit,
  TemplateRef,
  ViewContainerRef,
} from '@angular/core';
import { Subject } from 'rxjs';
import {
  distinctUntilChanged,
  filter,
  map,
```

```
  takeUntil,
  withLatestFrom,
} from 'rxjs/operators';

import { LocaleStateService } from '../../locale/data-
access/locale-state.service';

const directionQueryPattern = /^\(dir:
(?<direction>ltr|rtl)\)$/;

@Directive({
  exportAs: 'bidiMedia',
  selector: '[media]',
})
export class BidiMediaDirective implements OnDestroy,
OnInit {
```

2. In the following code, we can see a private observable property for the direction that has been applied to the application. This is calculated based on the selected application locale by using the `getLocaleDirection` function that we imported from the `@angular/common` package in the previous code.

   The destroy subject is used to manage subscriptions for the side effects registered by this directive.

   A value is emitted from the query direction subject every time a valid direction query is passed to this directive. The query direction observable prevents duplicate values from being emitted.

   The valid state observable emits a value as soon as an application direction and a direction query is made available to this directive.

   The `#view` field is used to hold a reference to the `<source>` element that this structural directive is applied to and conditionally renders it:

```
   #appDirection$ = this.localeState.locale$.pipe(
     map((locale) => getLocaleDirection(locale))
   );
   #destroy = new Subject<void>();
   #queryDirection = new Subject<Direction>();
   #queryDirection$ = this.#queryDirection.
   pipe(distinctUntilChanged());
```

```
#validState$ = this.#queryDirection$.pipe(
  withLatestFrom(this.#appDirection$),
  map(([queryDirection, appDirection]) => ({
  appDirection, queryDirection })),
  filter(
    ({ appDirection, queryDirection }) =>
      appDirection !== undefined && queryDirection !==
      undefined
  )
);
#view?: EmbeddedViewRef<HTMLSourceElement>;
```

3. The `media` input property validates the direction query and emits it through the query direction subject if it is valid.

   The template that is represented by this structural directive is injected into the constructor. The view container that we attach the template to when we conditionally render it is injected next.

   Finally, the locale state service is injected:

```
@Input()
set media(query: string) {
  if (!this.isDirection(query)) {
    throw new Error(
      `Invalid direction media query "${query}". Use
      format "(dir: ltr|rtl)"`
    );
  }

this.#queryDirection.next(this.queryToDirection(query));
}

constructor(
  private template: TemplateRef<HTMLSourceElement>,
  private container: ViewContainerRef,
  private localeState: LocaleStateService
) {}
```

4. Using the `OnInit` life cycle hook, we can register two side effects. The `OnInit` life cycle hook is used to manage subscriptions.

   We can use the `attachElement` method to attach the template when the direction query matches the application's direction:

```
ngOnInit(): void {
  this.attachElementOnDirectionMatch();
  this.removeElementOnDirectionMismatch();
}

ngOnDestroy(): void {
  this.#destroy.next();
  this.#destroy.complete();
}

private attachElement(): void {
  if (this.#view) {
    return;
  }

  this.#view = this.container.createEmbeddedView(this.
  template);
}

private attachElementOnDirectionMatch(): void {
  const directionMatch$ = this.#validState$.pipe(
    filter(
      ({ appDirection, queryDirection }) =>
      queryDirection === appDirection
    )
  );

  directionMatch$
    .pipe(takeUntil(this.#destroy))
    .subscribe(() => this.attachElement());
}
```

5. The `isDirection` method is used by the `media` input property we saw in a previous step.

   The `queryToDirection` method splits `query` into parts and extracts the direction value from it:

```
private isDirection(query: string): boolean {
  return directionQueryPattern.test(query);
}

private queryToDirection(query: string): Direction {
  const { groups: { direction } = {} } = query.
  match(directionQueryPattern)!;

  return direction as Direction;
}
```

6. The `removeElement` method is used to remove the elements that this directive is applied to whenever the direction query does not match the application's direction:

```
private removeElement(): void {
  this.container.clear();
}

private removeElementOnDirectionMismatch(): void {
  const directionMismatch$ = this.#validState$.pipe(
    filter(
      ({ appDirection, queryDirection }) =>
      queryDirection !== appDirection
    )
  );

  directionMismatch$
    .pipe(takeUntil(this.#destroy))
    .subscribe(() => this.removeElement());
}
}
```

The following screenshot shows what it looks like when combined with the locale picker and host direction service from the previous sections:

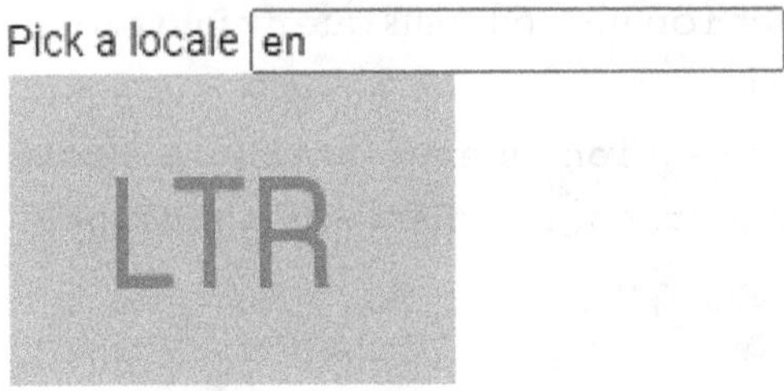

Figure 1.2 – Left-to-right demo image

In the following screenshot, we can see what an example layout looks like when a right-to-left direction language is picked using the locale picker:

Figure 1.3 – Right-to-left demo image

In this comprehensive section, we covered configuring Angular locales and lazy loading them. To demonstrate the power of lazy-loaded locales, we used our knowledge to create a directionality-aware structural directive for media content that can be used for art direction and combined it with a locale picker.

The next section explores changes and additions to Angular's testing APIs.

Enhancing tests with stronger types and new APIs

For many different types of tests, Angular's `TestBed` API is both necessary and useful. Ivy introduces a strongly typed API for resolving dependencies through the Angular testing module injector, which can be configured using the static `TestBed.configureTestingModule` method. In this section, we will explore stronger typing in Angular tests.

Let's also look at an integrated component test for an Angular component using a custom Angular Material SVG icon. This can be done using the `FakeMatIconRegistry` service that was introduced with Angular Ivy.

Resolving strongly typed dependencies with TestBed.inject

`TestBed.get` always returns a value of the `any` type. This static deprecated method has not been deprecated as of Angular version 12, but it could be removed in any major version following that. Its replacement is the type-safe `TestBed.inject` static method.

Let's look at a couple of simple examples to see the immediate difference in usage:

```
describe('inside of a test case', () => {
  it('TestBed.get does not infer the dependency type', () => {
    const service: MyService = TestBed.get(MyService);
  });

  it('TestBed.inject infers the dependency type', () => {
    const service = TestBed.inject(MyService);
  });
});
```

In both examples, the `service` variable has a `MyService` type. Notice that when using `TestBed.get`, we must declare the variable type ourselves. When using `TestBed.inject`, the type is inferred based on the value we pass to the static method.

When sharing variables between test cases, using `TestBed.get` can get us into trouble. Let's walk through an example to understand this issue:

```
describe('when sharing a variable between test cases', () => {
  decribe('TestBed.get does not identify dependency type
  issues', () => {
    let service: MyService;

    beforeEach(() => {
      service = TestBed.get(TheirService);
    });

    it('calls a service method', () => {
```

```
      service.myMethod();
    });
  });

  decribe('TestBed.get inject identifies dependency type
  issues', () => {
    let service: MyService;

    beforeEach(() => {
      service = TestBed.inject(TheirService);
    });

    it('calls a service method', () => {
      service.myMethod();
    });
  });
});
```

The test case with a setup hook using `TestBed.get` will only throw a runtime error when executing a test case that calls a method that does not exist on `TheirService`.

With the strongly typed `TestBed.inject`, the test case setup hook will throw a compile-time error, indicating that the interfaces of `MyService` and `TheirService` are incompatible. This is the power of TypeScript's static analysis at work.

Let's look at the signatures of both methods to get an overview of their parameters and return types:

```
type ProviderToken<T> = Type<T> |
InjectionToken<T> | AbstractType<T>;

class TestBed {
  static get(token: any, notFoundValue?: any): any;
  static get<T>(
    token: ProviderToken<T>,
    notFoundValue?: T,
    flags?: InjectFlags
  ): any;

```

```
  static inject<T>(
    token: ProviderToken<T>,
    notFoundValue?: T,
    flags?: InjectFlags
  ): T;
  static inject<T>(
    token: ProviderToken<T>,
    notFoundValue: null,
    flags?: InjectFlags
  ): T | null;
  static inject<T>(
    token: ProviderToken<T>,
    notFoundValue?: T | null,
    flags?: InjectFlags
  ): T | null;
}
```

Before `TestBed.get` was deprecated, the method signature similar to `TestBed.inject` was added to make room for an easier migration path. However, as we can see, both its method signatures return a value of the `any` type, which can lead to situations that we illustrated in the examples earlier in this section.

Let's walk through the parameters to enhance our understanding of `TestBed.inject`.

The `token` parameter accepts a value of one of three generic types described by the generic `ProviderToken<T>` union type. The first generic type is `Type<T>`, which means a concrete class type. The second generic type that's supported is `InjectionToken<T>`, which is the dependency injection token that we are all familiar with. The generic type parameter, `T`, of the passed injection token aids with type inference. The final type that the `token` parameter can have is `AbstractType<T>`, which refers to an abstract class type; this is a class that cannot be instantiated because of the `abstract` TypeScript modifier. An abstract class is useful for inheritance or as a *lightweight injection token*, as described in the Angular documentation.

The `notFoundValue` parameter is a bit special. We can leave it out to throw a runtime error if the dependency cannot be resolved. We can pass a default value of the `T` type to resolve that in case the dependency is not provided in the testing module injector. Finally, we can pass `null` combined with the `Optional` inject flag.

The optional `flags` parameter is a bitmask of zero or more `InjectFlags` binary values. In the case of the static `TestBed` methods for resolving dependencies, only the `InjectFlags.Optional` value is relevant and should be combined with passing `null` as an argument for the `notFoundValue` parameter.

There is one interesting thing to note here. `TestBed.inject` has the same method signature as the only non-deprecated signature of `Injector#get`. However, `Injector#get` also has a deprecated method signature, which allows any value, such as a text string or a number, to be used as a dependency injection token.

When `TestBed.get` is removed from Angular, we will be unable to work with deprecated dependency injection token types such as text strings, even though they might still be supported by `Injector#get`, until that method signature is eventually removed.

Now that we have learned about resolving strongly typed dependencies with `TestBed.inject`, let's move on to the `FakeMatIconRegistry`.

Stubbing custom Angular Material SVG icons with FakeMatIconRegistry

Angular Material's Icon component can be extended to support custom icons. In applications where we add custom SVG icons, we might run into an error when testing components using those custom SVG icons.

Ivy introduces `FakeMatIconRegistry` to replace the real `MatIconRegistry`, which normally keeps track of how to resolve static assets from icon names. To use it, import the Angular `MatIconTestingModule` module from the `@angular/material/icon/testing` subpackage.

Let's say that we have a spaceship launch button component, as shown in the following code:

```
@Component({
  changeDetectionStrategy: ChangeDetectionStrategy.OnPush,
  selector: 'spaceship-launch-button',
  template: `
    <button
      aria-label="Launch spaceship"
      color="accent"
      mat-fab
      (click)="onClick()"
    >
```

```
      <mat-icon svgIcon="myicons:spaceship"></mat-icon>
    </button>
  `,
})
export class SpaceshipLaunchButtonComponent {
  constructor(private spaceship: SpaceshipService) {}

  onClick(): void {
    this.spaceship.launch();
  }
}
```

Its integrated component test suite may look like this:

```
import { TestBed } from '@angular/core/testing';
import { MatIconModule } from '@angular/material/icon';
import { MatIconTestingModule } from '@angular/material/icon/
testing';
import { By } from '@angular/platform-browser';

import { SpaceshipLaunchButtonComponent } from './spaceship-
launch-button.component';

const MouseClickEvent = {
  Left: { button: 0 },
  Right: { button: 2 },
};

describe('SpaceshipLaunchButtonComponent', () => {
  let fixture:
  ComponentFixture<SpaceshipLaunchButtonComponent>;
  let serviceSpy: jasmine.SpyObj<SpaceshipService>;

  beforeEach( () => {
    serviceSpy = jasmine.
    createSpyObj<SpaceshipService>(SpaceshipService.name, [
      'launch',
    ]);
```

```
    TestBed.configureTestingModule({
      declarations: [SpaceshipLaunchButtonComponent],
      imports: [MatIconModule, MatIconTestingModule],
      providers: [{ provide: SpaceshipService, useValue:
      serviceSpy }],
    });
    TestBed.compileComponents();

    fixture = TestBed.
    createComponent(SpaceshipLaunchButtonComponent);
    fixture.detectChanges();
  });

  it('launches the spaceship when clicked', () => {
    const button = fixture.debugElement.query(By.
    css('button'));

    button.triggerEventHandler('click', MouseClickEvent.Left);

    expect(serviceSpy.launch).toHaveBeenCalledTimes(1);
  });
});
```

Notice that we imported `MatIconTestingModule` into the Angular testing module. This replaces `MatIconRegistry` with `FakeMatIconRegistry`, which returns an empty SVG when any SVG icon is requested. We still import `MatIconModule` to be able to render the icon button instead of using a lax compilation schema for a shallow component test.

Summary

In this first chapter, we discussed the modern language features that were introduced with the recent versions of TypeScript to accompany Angular Ivy. Through simple, common examples, we learned about nullish coalescing and optional chaining. We also identified the differences between TypeScript access modifiers and native private class members, all of which are features that we will make use of throughout the code in this book. By learning about these topics, you can now refactor your existing application or implement new features using these powerful language additions.

Globalization is needed for regional and multilingual support in Angular applications. We covered the basics of configuring multiple locales for our Angular build process. After that, we discovered our newfound ability to lazy load locale data such as regional number, currency, and date formats, as well as directionality information.

To demonstrate runtime locale switching, we created a locale picker and a direction service that set the direction of the application DOM based on the active locale. As an advanced example, we implemented a direction-aware media `<source>` element directive that can be used for art direction based on user locale.

We then discussed how the static `TestBed.inject` method is a type-safe replacement for `TestBed.get` that's used to resolve dependencies from the Angular testing module injector. We explored how strong typing can prevent mistakes when sharing variables between test steps.

Finally, we walked through an integrated component test for a component that used the `MatIcon` component. We imported `MatIconTestingModule` to use `FakeMatIconRegistry` for custom icons, allowing us to render the spaceship launch button component's template in its component test.

In the next chapter, we will discover new features that have been introduced by Angular Ivy in the tooling, configuration, and developer experience categories.

2

Boosting Developer Productivity Through Tooling, Configuration, and Convenience

Across data binding and attribute directives, there are many options to dynamically change styles in an Angular application. Angular Ivy supports multiple styling **application programming interfaces (APIs)** in a predictable way. This chapter teaches us about Angular Ivy's style binding precedence rules through a simple example that uses almost every possible style API in Angular.

Through a few examples, we will explore how directive and component inheritance and the sharing of metadata have changed in Angular Ivy. Before that, we learn which metadata properties are sharable through inheritance.

Ahead-of-time (**AOT**) compilation was introduced in one of the first major releases of Angular. Angular Ivy is the first generation to enable it in all phases of the application life cycle—from development through testing to builds. We will discuss how this affects our workflow and bundle size after briefly peeking under the hood of Angular to discuss some of the internal building blocks that have enabled the performance improvements needed to make this change.

The **strict mode** preset available with Angular Ivy enables additional compile-time checks of our Angular applications as well as bundle optimizations.

Angular Ivy and its supported versions of TypeScript have added significant speed improvements and stricter checks to tests. We will learn about this through examples that are easy to understand.

Angular's compiler analyzes our application code at build time. This is nice, but some of the compiler error messages suffer from missing context. In Angular Ivy, contextual details have been added to several compilation error messages. In this chapter, you will see Ivy compilation error examples compared to compilation errors from earlier generations of Angular as we discuss how it improves the developer experience.

Angular is known for great tooling and support for strong typing through TypeScript. However, we have not been able to type-check all parts of our component templates in the first or second generation of the Angular framework. The Ivy generation completes the component template type-checking story for the Angular framework.

Automated migration schematics are a wonderful piece of tooling that is part of the Angular framework. Since Ivy was released, they include messages with descriptions and an option to run them in separate commits, as we will see in this chapter.

The versions of the Angular Language Service introduced with Angular Ivy have allowed for better integration with **integrated development environments** (**IDEs**) such as **Visual Studio Code** (**VS Code**). Additional tooltips are added to component templates and additional details have been added to existing tooltips. Syntax highlighting has been added to inline templates and improved for external file templates.

In this second chapter, we are going to cover these topics:

- Using predictable style bindings
- Sharing metadata through directive and component inheritance
- Outputting faster and smaller bundles with AOT compilation
- Taking advantage of strict mode and other new configurations

- Enhancing our Angular testing experience
- Leveling up our developer experience

After finishing this chapter, you can boost your developer productivity by taking advantage of the latest improvements in tooling, configuration, and other Angular features.

Technical requirements

To support all features used in the code examples of this chapter, your application requires at least the following:

- Angular Ivy version 12.0
- TypeScript version 4.2

You can find complete code examples for component inheritance and style bindings in this book's companion GitHub repository at `https://github.com/PacktPublishing/Accelerating-Angular-Development-with-Ivy/tree/main/projects/chapter2`.

Using predictable style bindings

Angular has many ways to bind styles and classes to **Document Object Model** (**DOM**) elements. Ivy introduces predictable style bindings because of a precedence ruleset that covers all of Angular's style binding APIs except for the `NgClass` and `NgStyle` directives.

Template element bindings have higher priority than directive host bindings, which have higher priority than component host bindings. Binding of individual **Cascading Style Sheets** (**CSS**) classes and style properties have higher priority than binding maps of class names and style properties. Binding values that define the full `class` or `style` attributes have even lower priority. The `NgClass` and `NgStyle` directives override all other bindings on every value change.

Bottom values in style bindings are treated differently. Binding `undefined` will defer to lower-priority bindings, while `null` will override bindings with lower priority.

Let's look at the following example:

```
@Component({
  selector: 'app-root',
  template: `
```

```
    <app-host-binding
      [ngStyle]="{ background: 'pink' }"
      [style.background]="'red'"
      [style]="{ background: 'orange' }"
      style="background: yellow;"
      appHostBinding
    ></app-host-binding>
  `,
})
class AppComponent {}

@Directive({
  host: {
    '[style.background]': "'blue'",
    style: 'background: purple;',
  },
  selector: '[appHostBinding]',
})
class HostBindingDirective {}

@Component({
  host: {
    '[style.background]': "'gray'",
    style: 'background: green;',
  },
  selector: 'app-host-binding',
})
class HostBindingComponent {}
```

In the preceding code example, we see components and a directive using many different types of style bindings. Despite this, it will output only a single style rule to the DOM for the `<app-host-binding>` element. The background color of this rule will be evaluated as `pink`.

The order in which the background colors are applied is shown here, with the highest precedence first:

1. Pink (`NgStyle` directive binding)
2. Red (template property binding)
3. Orange (template map binding)
4. Yellow (static style value)
5. Blue (directive host property binding)
6. Purple (static directive host style binding)
7. Gray (component host property binding)
8. Green (static component host style binding)

As seen in the example, the order in which the bindings are mentioned in templates and metadata options does not matter—the precedence ruleset is always the same.

Having predictable style bindings makes it easier to implement complex use cases in our applications. It is worth mentioning that another reason for introducing this breaking change is that Ivy does not guarantee the order in which data bindings and directives are applied.

In this section, we witnessed the following styling precedence rules in effect, from highest priority to lowest:

1. Template property bindings
2. Template map bindings
3. Static template class and style values
4. Directive host property bindings
5. Directive host map bindings
6. Static directive host class and style bindings
7. Component host property bindings
8. Component host map bindings
9. Static component host class and style bindings

The order in which style bindings are listed in code only matters if two bindings share the same precedence, in which case the last one wins.

`NgClass` and `NgStyle` directive bindings override all other style bindings. They are the `!important` equivalents of Angular style bindings.

Now that we can predict how multiple style bindings and values affect our **user interface (UI)**, let's look at how we can use **class inheritance** to share directive and component metadata.

Sharing metadata through directive and component inheritance

Angular Ivy changes directive and component inheritance in a more explicit but predictable manner, which allows the bundle size and compilation speed to decrease.

When a base class is using any of the following Angular-specific features, it has to have a `Directive` or `Component` decorator applied:

- `Dependency` or `Attribute` injection
- `Input` or `Output` properties
- `HostBinding` or `HostListener` bindings
- `ViewChild`, `ViewChildren`, `ContentChild`, or `ContentChildren` queries

To support this, we can add a `Directive` decorator without any options. This conceptually works like an abstract directive and will throw a compile-time error if declared in an Angular module.

We could make the base class `abstract`, but that would cause us to have to extend it to test it, so it is a trade-off.

By extending base directives, we can inherit the `inputs`, `outputs`, `host`, and `queries` metadata options. Some of them will even be merged if declared both in the subclass and base class.

Components are able to inherit the same metadata options from their base class but are unable to inherit `styles` and `template` metadata. It is possible to refer to the same `styleUrls` and `templateUrl` though.

Let's write some example components that share behavior through a base class. First, we will create a base search component, as seen in the following code snippet:

```
import { Component, EventEmitter, Input, Output } from '@
angular/core';
import { debounceTime, distinctUntilChanged } from 'rxjs/
operators';

@Component({
```

```
  selector: 'app-base-search',
  template: '',
})
export class BaseSearchComponent {
  #search = new EventEmitter<string>();

  @Input()
  placeholder = 'Search...';
  @Output()
  search = this.#search.pipe(debounceTime(150),
   distinctUntilChanged());

  onSearch(inputEvent: Event): void {
    const query = (inputEvent.target as
     HTMLInputElement)?.value;

    if (query == null) {
      return;
    }

    this.#search.next(query);
  }
}
```

The base search component has an event handler that handles input events representing a search query. It debounces searches for `150` **milliseconds (ms)** and ignores duplicate search queries before outputting them through its `search` output property. Additionally, it has a `placeholder` input property.

Next, we will create a simple search box component that inherits from the base search component, as seen in the following code snippet:

```
import { Component } from '@angular/core';

import { BaseSearchComponent } from './base-search.component';

@Component({
  selector: 'app-search-box',
```

```
  styleUrls: ['./base-search.scss'],
  template: `
    <input
      type="search"
      [placeholder]="placeholder"
      (input)="onSearch($event)"
    />
  `,
})
export class SearchBoxComponent extends BaseSearchComponent {}
```

The search box component uses base search styles and can add its own component-specific styles if it needs to. The `<input>` element in its component template binds to the `placeholder` input property it inherits from the base search component. Likewise, the `input` event is bound to the `onSearch` event handler it inherits.

Let's create another component that inherits from the base search component. The following code block lists the suggested search component:

```
import { Component, Input } from '@angular/core';

import { BaseSearchComponent } from './base-search.component';

@Component({
  selector: 'app-suggested-search',
  styleUrls: ['./base-search.scss'],
  template: `
    <input
      list="search-suggestions"
      [placeholder]="placeholder"
      (input)="onSearch($event)"
    />
    <datalist id="search-suggestions">
      <option *ngFor="let suggestion of suggestions"
        [value]="suggestion">
        {{ suggestion }}
      </option>
    </datalist>
```

```
  `,
})
export class SuggestedSearchComponent extends
BaseSearchComponent {
  @Input()
  suggestions: readonly string[] = [];
}
```

In addition to the inherited input property, `placeholder`, the suggested search component adds a `suggestion` input property, which is a list of search query suggestions. The component template loops over these suggestions and lists them as `<option>` elements in a `<datalist>` element that is tied to the `<input>` element.

Similar to the search box component, the suggested search component binds to the `onSearch` event handler and the `placeholder` input property. It also uses the base search styles.

> **Important Note**
> As seen in these examples, we do not have to add duplicate constructors in subclasses to enable constructor injection.

Directive and component metadata is the special glue that ties TypeScript classes to the DOM through component templates and data binding. Through classical **object-oriented programming** (**OOP**) patterns, we are familiar with sharing properties and methods through class inheritance.

In this section, we learned how Ivy has enabled us to share metadata in a similar, predictable way through metadata-enabled class inheritance.

The next topic we are going to explore is AOT compilation. Ivy heralds the era of AOT compilation everywhere in Angular applications.

Outputting faster and smaller bundles with AOT compilation

Angular Ivy is first and foremost an internal rewrite of the Angular compiler, the Angular runtime, and a few more pieces of the framework. A lot of effort was put into maintaining backward compatibility with application code written for the View Engine generation of the Angular framework (versions 4 through 8).

At the heart of this new generation is the **Ivy Instruction Set**, which is a runtime DOM instruction set similar to the **Incremental DOM** library by Google. In a nutshell, an Angular component is compiled into two lists of instructions that, when executed by Ivy, will initialize the DOM with the first list of instructions and update the DOM with the second list of instructions whenever changes are detected.

The Ivy Instruction Set is lighter than the View Engine equivalent, which the runtime had to translate before applying it. In View Engine, it used to be the case that a large Angular application would eventually have a smaller **just-in-time** (**JIT**)-compiled bundle than the AOT-compiled bundle for the same application. With Ivy, this is no longer the case.

For small Angular applications, Ivy adds the benefit that the Ivy Instruction Set and the individual parts of the runtime are tree-shakable. This means that we are not paying for the parts of the framework runtime that we are not using, in that they are removed from the output bundle—for example, Angular's animation and globalization APIs are not part of our application bundle unless we are using them. This is very important in **microfrontends** and **web components**, where we need to minimize the overhead of the Angular framework.

Angular version 11 is the final major version supporting View Engine. The Ivy Instruction Set is kept as a code-generation API, meaning that our code should not rely on it. Angular libraries supporting Angular versions 9-11 are not recommended to ship Ivy-compiled code. Instead, the Angular Compatibility Compiler is used to transpile View Engine-compiled libraries to Ivy. Unfortunately, this means that our initial build time increases as the Angular Compatibility Compiler must compile each Angular library entry point in our `node_modules` folder. This is especially hard to manage in managed build pipelines.

Read about the Angular Compatibility Compiler in *Chapter 10, Using the Angular Compatibility Compiler.*

As of Angular version 12, it is recommended for Angular libraries to ship partially Ivy-compiled bundles using the Ivy compiler. This means that our application can more fully enjoy incremental builds that speed up the overall build time. This is because of an Ivy concept known as the **principle of locality**. The Ivy compiler only needs to know about the public API of, for example, child components rather than having to know about their injected dependencies and mention them in the compiled output of the parent component.

In general, Ivy results in decreased build time, except for the time it takes to run the Angular Compatibility Compiler on Angular libraries, including the Angular framework's packages.

As we discussed, using the Ivy compiler also results in decreased bundle sizes overall. However, if we are using most parts of the framework runtime, our main chunk's bundle size might increase while the size of lazy-loaded chunks decreases.

With the introduction of Angular Ivy, we can now use AOT compilation during all phases of development. This removes subtle but important differences between development, testing, the build process, and production runtime.

The short explanation in this section is just enough to give us an overall idea about the effect that AOT compilation has on Angular applications. However, this book is a practical approach to developing Angular Ivy applications. In everyday development, we do not have to be familiar with the internals of the Angular compiler, rendering engine, and runtime.

Read more about the impact and limitations of Angular's AOT compiler in *Chapter 12, Embracing Ahead-of-Time Compilation.*

In the following section, we look at the compilation presets and configurations introduced by Ivy and its accompanying TypeScript versions.

Taking advantage of strict mode and other new configurations

Angular has always been big on tooling. Ivy adds and enables additional configurations that help us catch errors early and output smaller bundles.

Strict mode

When we create an Angular workspace using the `ng new` command, the `--strict` parameter flag is on by default as of Angular version 12. Using the `strict` workspace preset enables additional static analysis. The `--strict` parameter flag is also supported for project generation schematics such as `ng generate application` and `ng generate library`.

The strict preset sets the following TypeScript compiler options to `true`:

- `forceConsistentCasingInFileNames`
- `noImplicitReturns`
- `noFallthroughCasesInSwitch`
- `strict`

These options help us catch a lot of potential errors at compile time. Refer to the TypeScript documentation for details about the individual compiler options. The `strict` TypeScript compiler option is a shorthand for enabling all the following compiler options as of TypeScript version 4.2:

- `alwaysStrict`
- `noImplicitAny`
- `noImplicitThis`
- `strictBindCallApply`
- `strictFunctionTypes`
- `strictNullChecks`
- `strictPropertyInitialization`

Additionally, future strict TypeScript compiler options will be enabled automatically when using the `strict` shorthand.

Bundle budgets are reduced in strict mode. The initial bundle warns at 500 **kilobytes** (**KB**) and errs at 1 **megabyte** (**MB**), while component styles warn at 2 KB and err at 4 KB.

Strict mode further enables strict template type checking, which we will cover later in this chapter.

Angular compiler options

Only a few Angular compiler option defaults have changed in Ivy. In Angular version 9, the default value of the `enableIvy` option was set to `true` for applications but `false` for libraries. This changed in Angular version 12, where the `enableIvy` option was removed entirely when View Engine support was disabled for Angular applications and the `"partial"` value for the `compilationMode` option was added for partial Ivy compilation of Angular libraries.

The `strictTemplates` option is introduced with Ivy. Its default value is `false`, but it is set to `true` when generating an Angular workspace by using the `ng new` command, which enables the `--strict` parameter flag. The same applies to the `fullTemplateTypeCheck` option, which is implicitly set by the `strictTemplates` option.

The default value of the `strictInjectionParameters` and `strictInputAccessModifiers` options is still `false`, but it is set to `true` when generating an Angular workspace by using the `ng new` command.

Now that you are aware of the many helpful configuration options introduced by Angular Ivy, you can gain more confidence about how your Angular application will behave in production runtime.

Next up is a favorite topic of ours: testing. Both Ivy and its related TypeScript versions introduce substantial improvements to the Angular testing experience.

Enhancing our Angular testing experience

Ivy is a major milestone for Angular tests. Besides the stronger typing in tests discussed in *Chapter 1, Discovering New APIs and Language Syntax,* it adds major speed improvements and useful test utilities, one of which is component testing harnesses, which we will cover in *Chapter 4, Exploring Angular Components Features.*

In this section, we explore how we can introduce values of unexpected types using a TypeScript annotation, which proves to be useful in tests. After that, we discuss another important aspect of AOT compilation in Angular Ivy.

Expect error compiler annotation

TypeScript version 3.9 introduces a special compiler instruction comment that is useful in tests.

The `@ts-expect-error` annotation comment allows values of incompatible types to be passed to functions in the following statement. As an example, let's write an `add` function and verify that it rejects strings—even at runtime—for robustness:

```
function add(left: number, right: number): number {
  assertIsNumber(left);
  assertIsNumber(right);

  return left + right;
}
```

The robust `add` function in the previous code snippet applies an assertion function for both operands. Let's test that an error is thrown if strings are passed, as follows:

```
describe('add', () => {
  it('rejects a string as left operand', () => {
    const textInput = '2';
    const four = 4;

```

```
    // @ts-expect-error
    expect(() => add(textInput, four)).toThrow();
  });

  it('rejects a string as right operand', () => {
    const three = 3;
    const textInput = '5';

    // @ts-expect-error
    expect(() => add(three, textInput)).toThrow();
  });
});
```

If we remove the `@ts-expect-error` comments, the TypeScript compiler throws errors because of the incompatible values we pass in the tests in the previous code block.

How is this different from `@ts-ignore` comments? The `@ts-expect-error` comments warn us if a compilation error is not thrown in the statement that follows. This raises our confidence in the code.

Faster tests with AOT compilation

Angular Ivy introduces AOT compilation to tests. This makes the test environment close to the production environment, which is a good trait as it allows us to catch errors early.

Until Ivy, Angular had a long-standing issue of relatively slow tests when they involved component tests using `TestBed`. The tests were slow because the test runner was reading, parsing, and compiling one or more files for every component per test case, not per test suite or per test run. Ivy introduces the principle of locality as well as a cache for compiled declarables and Angular modules, which speeds up component tests significantly. Additionally, rebuilds are faster, which improves speed when writing tests and fixing bugs.

With these pieces of valuable information, you now know how Ivy can greatly impact your developer workflow when implementing unit tests. As mentioned in the introduction of this section, we have saved one of the most exciting features for *Chapter 4, Exploring Angular Components Features*—namely, component testing harnesses.

The next section is all about how Ivy boosts our productivity by improving the Angular developer experience.

Leveling up our developer experience

Ivy improves the developer experience in many ways. In this section, we learn about the most noteworthy of them.

Improved compilation errors

Compile-time errors are a positive side effect of Angular using an AOT compiler. However, some of the build errors have needed additional context to aid in pinpointing the source of the error.

As an example, look at the error message in the following listing that is output when using an unknown element in an Angular application using View Engine:

```
ERROR in 'app-header' is not a known element:
1. If 'app-header' is an Angular component, then verify that it
is part of this module.
2. If 'app-header' is a Web Component then add 'CUSTOM_
ELEMENTS_SCHEMA' to the '@NgModule.schemas' of this component
to suppress this message. ("[ERROR ->]<app-header></app-header>

<p>My content</p>
")
```

Which component has the error? How do we fix it?

The next listing shows the error message that is output for the same mistake in an Angular application using Ivy:

```
ERROR in src/app/app.component.html:1:1 - error NG8001:
'app-header' is not a known element:
1. If 'app-header' is an Angular component, then verify that it
is part of this module.
2. If 'app-header' is a Web Component then add 'CUSTOM_
ELEMENTS_SCHEMA' to the '@NgModule.schemas' of this component
to suppress this message.

1 <app-header></app-header>
  ~~~~~~~~~~~~~~~~~~~~~~~~~

  src/app/app.component.ts:5:16
    5   templateUrl: './app.component.html',
```

```
                    ~~~~~~~~~~~~~~~~~~~~~~~~
      Error occurs in the template of component AppComponent.
```

The file path of the component model and the component template files are both listed, as well as the line number and the content of those lines. It is now much clearer where the problem was encountered—a big win for developer productivity.

Let's look at another example. If we accidentally add a duplicate comma to the `imports` array of an Angular module, as seen in the following code block, a build error will be output:

```
import { CommonModule } from '@angular/common';
import { NgModule } from '@angular/core';
import { BrowserModule } from '@angular/platform-browser';

import { AppComponent } from './app.component';

@NgModule({
  declarations: [AppComponent],
  imports: [BrowserModule, , CommonModule],
  providers: [],
  bootstrap: [AppComponent],
})
export class AppModule {}
```

View Engine outputs an error message similar to the one shown in the following listing:

```
ERROR in src/app/app.module.ts(11,27): Error during template
compile of 'AppModule'
  Expression form not supported.
src/app/app.module.ts(11,27): Error during template compile of
'AppModule'
  Expression form not supported.
Cannot determine the module for class AppComponent in C:/
projects/sandbox/angular-cli8-app/src/app/app.component.ts! Add
AppComponent to the NgModule to fix it.
```

We get the relevant file and line number, but the error description, `Expression form not supported`, is not very helpful. Here is the equivalent error message output by Ivy:

```
ERROR in projects/second-app/src/app/app.module.ts:11:3 - error
TS2322: Type '(typeof CommonModule | typeof BrowserModule |
undefined)[]' is not assignable to type '(any[] | Type<any> |
ModuleWithProviders<{}>)[]'.
  Type 'typeof CommonModule | typeof BrowserModule |
undefined' is not assignable to type 'any[] | Type<any> |
ModuleWithProviders<{}>'.
    Type 'undefined' is not assignable to type 'any[] |
Type<any> | ModuleWithProviders<{}>'.

11   imports: [BrowserModule, , CommonModule],
     ~~~~~~~
```

In the previous listing, we immediately see that there is an issue in the `imports` array and that somehow, an `undefined` value ended up in it. Context is provided and the error description is improved.

Finally, let's look at the improved description of an error triggered by a piece of code that is not statically analyzable, as follows:

```
import { Component } from '@angular/core';

const template = location.href;

@Component({
  selector: 'app-root',
  styleUrls: ['./app.component.css'],
  template,
})
export class AppComponent {}
```

The `template` option in the previous example is not statically determinable as it relies on information only available at runtime. Here is the error message output by View Engine:

```
ERROR in No template specified for component AppComponent
```

In the previous listing, we see that the View Engine error message is for metadata that is not statically determinable. It tells us which component has an error, but it does not help us understand how to fix it. The type information shows that `location.href` is of type `string`, so why does `AppComponent` not have a `template` option according to the compiler?

The error message output by Ivy is a lot more helpful, as seen in the following listing:

```
ERROR in src/app/app.component.ts:8:3 - error NG1010: template
must be a string
  Value could not be determined statically.

8   template,
    ~~~~~~~~

  src/app/app.component.ts:3:18
    3 const template = location.href;
                       ~~~~~~~~~~~~~
    Unable to evaluate this expression statically.
  node_modules/typescript/lib/lib.dom.d.ts:19441:13
    19441 declare var location: Location;
                      ~~~~~~~~~~~~~~~~~~
    A value for 'location' cannot be determined statically, as
it is an external declaration.
src/app/app.module.ts:8:5 - error NG6001: The class
'AppComponent' is listed in the declarations of the NgModule
'AppModule', but is not a directive, a component, or a pipe.
Either remove it from the NgModule's declarations, or add an
appropriate Angular decorator.

8     AppComponent
      ~~~~~~~~~~~~

  src/app/app.component.ts:10:14
    10 export class AppComponent { }
                    ~~~~~~~~~~~~
    'AppComponent' is declared here.
```

The error message output by Ivy doesn't just show us the line and expression that are not statically determinable. It goes one step further and shows us that the entire `location` object is non-deterministic at compile time.

Read more about AOT limitations in *Chapter 12, Embracing Ahead-of-Time Compilation.*

Strict template type checking

View Engine had basic and full modes for template type checking. Ivy introduces strict template type checking. In addition to the full-mode template type checks, strict mode enables several checks, outlined as follows:

- Property binding types are checked against their corresponding input property type.
- Property binding type checks are strict about `null` and `undefined` values.
- Generic types for components and directives are inferred and checked.
- Checks the type of template context variables, including `$implicit`.
- Checks the type of the `$event` template reference.
- Checks the type of template references to DOM elements.
- Safe navigation operations are type-checked.
- Array and object literals in component templates are type-checked.
- Attribute bindings are type-checked.

The strict check of `null` for property bindings is important for property bindings using `AsyncPipe` as it initially emits a `null` value. This means that input properties being used with `AsyncPipe` must either have a type that includes `null` or use the concepts known as template guards and input setter type hints.

An even better update experience

As part of Ivy, the Angular **command-line interface** (**CLI**) adds three improvements to the update experience.

When running the `ng update` command, the Angular CLI first downloads the latest stable version of the Angular CLI and uses it for the update to take advantage of the most recent improvements.

As part of the `ng update` command, automated migrations are run by the Angular CLI. For every migration, a message is shown, such as the example shown in the following listing:

```
** Executing migrations of package '@angular/core' **

> Static flag migration.
  Removes the `static` flag from dynamic queries.
  As of Angular 9, the "static" flag defaults to false and is
no longer required for your view and content queries.
  Read more about this here: https://v9.angular.io/guide/
migration-dynamic-flag
  Migration completed.
```

Finally, the `--create-commits` parameter flag is introduced to create a Git commit per migration to make it easier to debug the update process.

We will cover the Angular update process in more detail in *Chapter 11, Migrating Your Angular Application from View Engine to Ivy.*

Better IDE integration

Ivy introduces remarkable improvements to the Angular Language Service, which integrates with our IDE, such as VS Code.

Template and style **Uniform Resource Locators** (**URLs**) are verified inline, as seen in the following screenshot, where an error is displayed. This is especially helpful when renaming components:

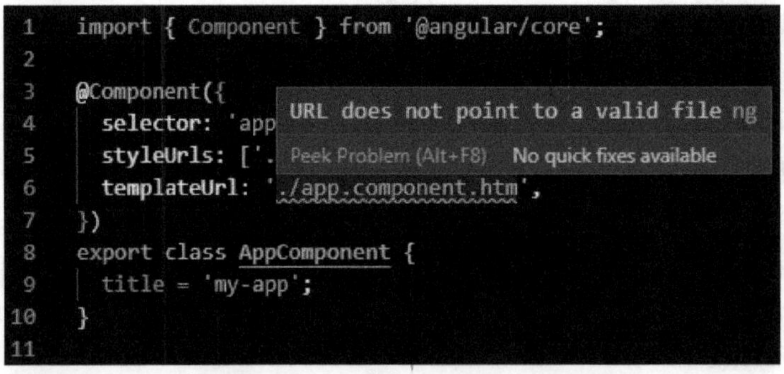

Figure 2.1 – Invalid template URL error

Another neat improvement is that we can use the *Go to definition* IDE command to navigate to a component template or style sheet file. This makes it easier to work with one specific component.

Ivy versions of the Angular Language Service give us an additional context in tooltips for component models. The following screenshot shows a tooltip for a child component element in a component template. In the tooltip, we can see the Angular module that declared the child component. In the example, we see that `ChildComponent` is declared by `AppModule`:

```
(component) AppModule.ChildComponent: typeof ChildComponent
<app-child></app-child>
```

Figure 2.2 – NgModule annotation in the component element tooltip

Similarly, the tooltip that appears for a component model also shows the Angular module that declared this component, as seen in the next screenshot. Here, we see that `AppComponent` is declared by `AppModule`:

```
import { Component } from '@angular/core';

@Component({
  selector: 'app-root',
  styleUrls:   (component) AppModule.AppComponent: class
  templateUrl
})             class AppComponent
export class AppComponent {
  title = 'my-app';
}
```

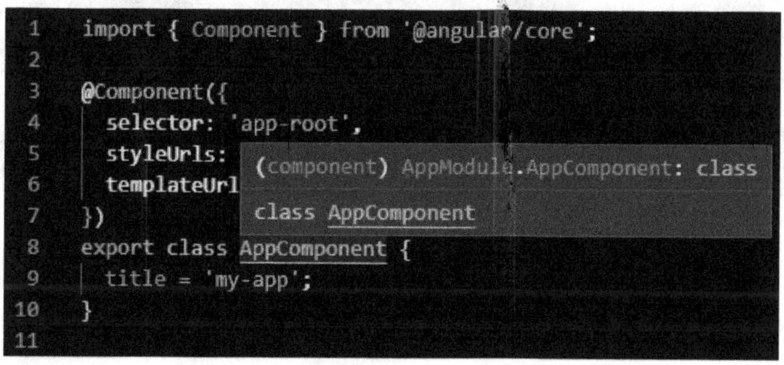

Figure 2.3 – NgModule annotation in the component model tooltip

With Ivy, we can see method signatures in a tooltip, as seen in the following screenshot. This helps us reason about event handlers:

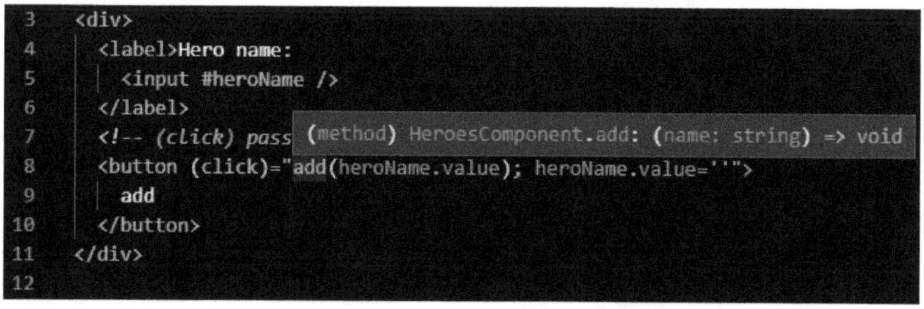

```
<div>
  <label>Hero name:
    <input #heroName />
  </label>
  <!-- (click) pass (method) HeroesComponent.add: (name: string) => void
  <button (click)="add(heroName.value); heroName.value=''">
    add
  </button>
</div>
```

Figure 2.4 – Component method signature in the tooltip

Ivy's Angular Language Service enables us to see type annotations of UI properties and the `$implicit` template reference. This allows us to get type information about an iterable UI property used with the `NgFor` directive, as seen in the following screenshot:

```
<ul class="heroes">
  <li *ngFor="let hero of her (variable) hero: Hero
    <a routerLink="/detail/{{hero.id}}">
      <span class="badge">{{hero.id}}</span> {{hero.name}}
    </a>
    <button class="delete" title="delete hero" (click)="delete(hero)">x</button>
  </li>
</ul>
```

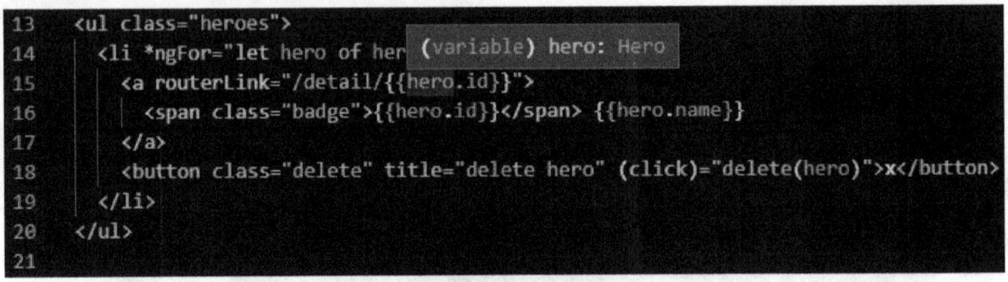

Figure 2.5 – Iterable UI property tooltip

The next screenshot shows how Ivy enables us to get type information about the named iterator template reference in each loop cycle:

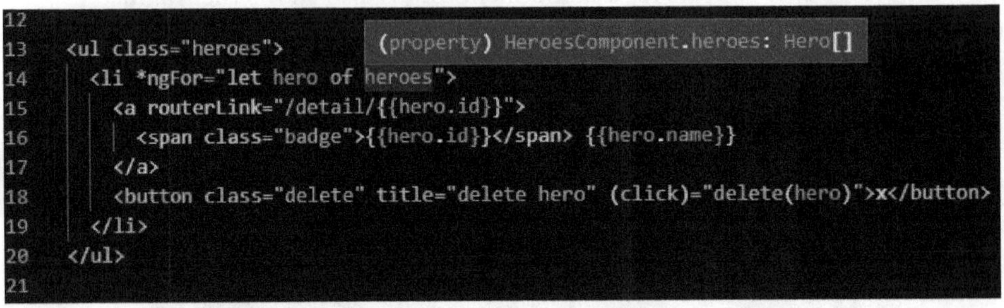

```
                     (property) HeroesComponent.heroes: Hero[]
<ul class="heroes">
  <li *ngFor="let hero of heroes">
    <a routerLink="/detail/{{hero.id}}">
      <span class="badge">{{hero.id}}</span> {{hero.name}}
    </a>
    <button class="delete" title="delete hero" (click)="delete(hero)">x</button>
  </li>
</ul>
```

Figure 2.6 – Named iterator template reference tooltip

Combined with strict template type checking, as described earlier in this chapter, this allows us to catch type errors in components early and consistently.

The improved Angular Language Service released with Ivy adds syntax highlighting to inline component templates and styles, as seen in the following screenshot. This makes single-file components easier to use:

```
@Component({
  selector: 'app-hero-detail',
  template: `
    <div *ngIf="hero">
      <h2>{{hero.name | uppercase}} Details</h2>
      <div><span>id: </span>{{hero.id}}</div>
      <div>
        <label>name:
          <input [(ngModel)]="hero.name" placeholder="name"/>
        </label>
      </div>
      <button (click)="goBack()">go back</button>
      <button (click)="save()">save</button>
    </div>
  `,
  styles: [`
    /* HeroDetailComponent's private CSS styles */
    label {
      display: inline-block;
      width: 3em;
      margin: .5em 0;
      color: #607D8B;
      font-weight: bold;
    }
```

Figure 2.7 – Syntax highlighting for inline component template and styles

Inline templates even benefit from the ability to add syntax highlighting to template expressions also introduced with Ivy.

Finally, Ivy adds support for style preprocessors such as Sass in inline styles by introducing the `inlineStyleLanguage` option to the Angular `@angular-devkit/build-angular:browser` and `@angular-devkit/build-angular:karma` builders, for example: `"inlineStyleLanguage": "scss"`.

Summary

In this chapter, we learned how Angular Ivy boosts our developer productivity through even better tooling and predictable APIs. We started out by learning the many ways of binding styles to elements using Angular Ivy. Through a comprehensive example, we saw how the order of style bindings does not matter. However, we learned that the type of style binding matters as it follows a precedence or priority defined by Angular Ivy. This makes the resolution of multiple style bindings predictable, which is important for implementing certain complex use cases. Finally, we discussed that this change was necessary because Ivy does not guarantee the order in which directives and data bindings are applied.

Ivy requires `Directive` and `Component` decorators on base classes that rely on Angular features, such as input and output properties or child queries. We learned how this results in a pattern that makes it easy to share metadata through inheritance in a predictable way.

To learn how Ivy's AOT compiler is faster and, for certain Angular projects, produces smaller bundles, we discussed how a lot of the internals were rewritten between View Engine and Ivy. We learned how the Ivy Instruction Set is a good fit for tree shaking to remove the parts of the framework that are unused by our applications.

With this chapter completed, you now know how Angular's strict mode preset and other configurations allow us to catch errors sooner, resulting in a more robust code base.

We discussed how Angular Ivy significantly improves the testing experience as it introduces major speed improvements and useful test utilities. We saw examples of how TypeScript allows us to expect a type error in tests to create APIs that are more robust at runtime. We discussed how AOT compilation is enabled in tests to reduce the gap between tests and runtime. This can significantly improve developer feedback when implementing tests and application code.

Not only did Ivy improve our testing experience, but it is also a productivity booster when it comes to the overall developer experience. This chapter compared compiler error messages from View Engine to the same type of error messages in Ivy to demonstrate the additional context they include. We discussed how strict template type checking finalizes type checking in component templates.

We briefly discussed how Ivy improves the update experience through messages that are output during automated migrations and how we can use the `--create-commits` parameter flag to split the automated migrations into individual Git commits.

We ended by discussing most of the improvements in Angular's IDE integrations, such as the new invalid template and style URL errors, tooltips displaying and declaring Angular modules, and syntax highlighting for inline component templates and styles.

Chapter 3, Introducing CSS Custom Properties and New Provider Scopes, and *Chapter 4, Exploring Angular Components Features,* are going to prepare you for the Ivy features you are going to need to implement features on an existing Angular application in *Part 2,* Build a Real-World Application with the Angular Ivy Features You Learned. We will discuss topics such as CSS Custom Properties, the Clipboard API, component testing harnesses, and the platform provider scope, all in the context of Angular.

3
Introducing CSS Custom Properties and New Provider Scopes

In *Part 2, Build a Real-World Application with the Angular Ivy Features You Learned,* we are going to implement features for an existing application. To prepare for this challenge, we are going to discuss some of the most interesting features introduced by Angular Ivy.

CSS Custom Properties are browser-native CSS variables that can be changed at runtime. In this chapter, we will use simple examples to uncover their scoped nature when applied to the **Document Object Model (DOM)**. We will combine them with the power of Angular to show off some neat inspirational tricks.

Dependency injection is a powerful feature at the very core of the Angular framework. As Angular developers, we have come to appreciate the `root` provider shorthand, which declares an application-wide singleton dependency that is shared throughout an Angular application.

Angular Ivy introduces two more provider shorthands that declare the `any` and `platform` provider scopes. First, we will discuss the nature of the `any` provider scope and the use cases it is relevant to. Then, we will implement most of a feature using it, step by step, to get a deeper understanding of this advanced topic.

Finally, we will discuss the `platform` provider scope, which is easier to understand than the `any` provider scope but only applies to a few very specific use cases, namely Angular Elements-powered **web components** and Angular-based microfrontends.

In this chapter, we will cover the following topics:

- Using CSS Custom Properties with Angular
- Discovering the `platform` provider scope
- Understanding the `any` provider scope

These topics, combined with the knowledge you gained in the previous chapters, will give you the foundation you will need to implement features in an existing application in *Part 2 , Build a Real-World Application with the Angular Ivy Features You Learned,*.

Technical requirements

To support all the features that will be used in the code examples of this chapter, your application requires at least the following:

- Angular Ivy version 9.0
- TypeScript version 3.6

You can find the complete code examples for colored panels, the welcome banner, CSS Custom Properties, and the `any` provider scope in this book's companion GitHub repository at `https://github.com/PacktPublishing/Accelerating-Angular-Development-with-Ivy/tree/main/projects/chapter3`.

Using CSS Custom Properties with Angular

CSS Custom Properties are native, runtime CSS variables. They can be scoped to a sub-tree of the DOM, such as Angular's element injector hierarchy. We are used to doing this with CSS classes with increased specificity or CSS source ordering.

CSS Custom Properties do not rely on specificity or CSS source ordering. Instead, we can dynamically change the value at runtime or override it for a specific part of the DOM.

Let's look at a simple example. Here, we have listed both a global and a scoped declaration of the `--background-color` CSS custom property:

```
:root {
  --background-color: rebeccapurple;
}

.contrast {
  --background-color: hotpink;
}

.panel {
  background-color: var(--background-color);
}
```

Now, look at the following HTML and try to reason about the global and scoped CSS Custom Properties. Don't worry if you do not understand it immediately. We will discuss it in detail in a moment:

```
<div class="panel">
  This panel is purple.
</div>

<div class="panel contrast">
  This panel is pink.
</div>

<div class="contrast">
  This element is transparent.

  <div class="panel">
    This panel is also pink.
  </div>
</div>
```

The previous example demonstrates that the custom property value that was declared in the `:root` scope applies to all the DOM elements using the `--background-color` custom property. The first panel is purple, as declared in the root scope.

> **Important Note**
>
> A CSS custom property name must begin with two dashes (`--`). The name is case-sensitive.

The following screenshot shows the result, which is also available in this book's companion GitHub repository, as mentioned in the introduction to this chapter:

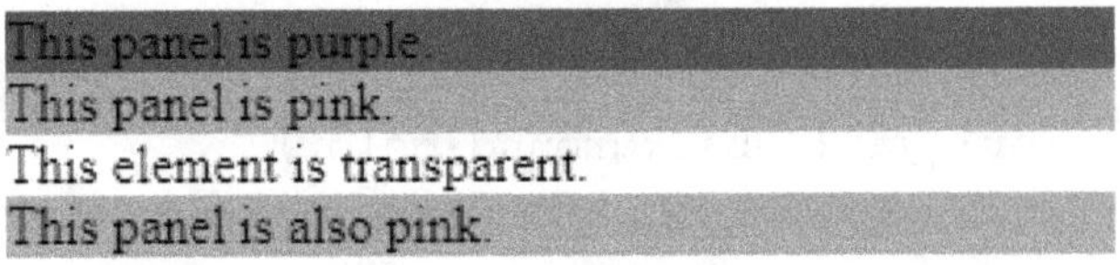

This panel is purple.
This panel is pink.
This element is transparent.
This panel is also pink.

Figure 3.1 – Scoped CSS Custom Properties in effect

As we can see, the second panel is pink. This is because the `contrast` CSS class declares a scope in which the `--background-color` custom property is pink instead of purple. The custom property scope starts at the DOM element where the class is used and affects all its descendants, as shown in the third panel. However, the element that's wrapping the third panel is neither purple nor pink since it does not have a class that uses the custom property, only one that declares its value.

You might be wondering what we can use CSS Custom Properties for. We can use them for anything that CSS can do, such as animations, transitions, fonts, colors, dimensions, and layouts.

They can hold any CSS value, including strings and numbers. For example, custom properties might be used for globalization to alternate text. We can do this using the `::before` or `::after` pseudoelements with the `content` property, combined with the `:lang()` pseudoclass. We will demonstrate this in a small example. This can be found in this book's companion GitHub repository:

1. In our global stylesheet, add the following rules, which set a welcome text CSS custom property based on the active language in a page or element:

```
:lang(en) {
  --welcome-text: 'Welcome to my website!';
}
```

```
:lang(fr) {
  --welcome-text: 'Bienvenue sur mon site web!';
}
```

2. Let's create a welcome component and add the following scoped styles to it. The element with the `welcome-content` class will contain the value of the CSS `--welcome-text` custom property. Updating the value of the custom property changes the content of this element:

```
.welcome-banner > .welcome-content::before {
  display: inline-block;
  content: '' var(--welcome-text) '';
}
```

3. The template of the welcome component contains buttons for selecting English and French. Clicking them triggers an event handler in the welcome component. The template also contains a welcome banner and an inline element for the welcome content. This element has its `lang` property bound to the `language` UI property of the welcome component:

```
<button (click)="onLanguageSelect('en')">EN</button>
<button (click)="onLanguageSelect('fr')">FR</button>

<p class="welcome-banner">
  <span class="welcome-content"
    [lang]="language"></span>
</p>
```

4. The welcome component model contains a language property – set to English by default – and an event handler for language selection:

```
import { Component } from '@angular/core';

@Component({
  selector: 'app-welcome',
  // Template and styles as listed above
})
export class WelcomeComponent {
  language = 'en';

```

```
    onLanguageSelect(language: string): void {
      this.language = language;
    }
  }
```

5. When clicking one of the welcome component's language buttons, the specified language is activated for the welcome content element, which changes the scoped value of the `--welcome-text` custom property. This, in turn, causes the welcome content to change.

The following screenshot shows the welcome banner in English, which is the default language:

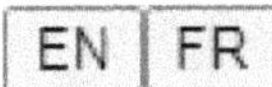

Welcome to my website!

Figure 3.2 – English welcome banner

When the French language button is clicked, the welcome content changes, as shown in the following screenshot:

EN FR

Bienvenue sur mon site web!

Figure 3.3 – French welcome banner

Take a minute to imagine how powerful combining Angular with dynamic custom properties can be. Angular Ivy introduces support for binding values to CSS Custom Properties by using a `style` property binding or host binding.

For example, to bind a value to a `--text-size` custom property, we can create a host binding in our component model, as shown in the following code:

```
import { Component, HostBinding } from '@angular/core';

@Component({
  selector: 'app-root',
  template: `
    <label>
      Text size
```

```
      <input type="number" min="10" max="48" step="2"
        [(ngModel)]="textSize" />
    </label>
    px

    <p>
      Lorem ipsum dolor sit amet, consectetur adipiscing
        elit. Donec accumsan, nisi sed aliquet lobortis, est
        lorem euismod libero, at rutrum lacus tellus non
        metus. Proin a nunc a libero vehicula egestas
        pretium id ipsum.
    </p>
  `,
})
export class AppComponent {
  @HostBinding('style.--text-size.px') textSize = 16;
}
```

We can use the root component's host element as the scope for our CSS Custom Properties:

```
p {
  font-size: var(--text-size, 16px);
}
```

The previous styles are declared in the global stylesheet to apply the text size that was picked to all paragraphs.

> **Important Note**
>
> Dialogs and other top-level DOM elements are mostly rendered outside of the root component's DOM subtree. We would also have to add host bindings to them or their wrapping elements, but this is outside the scope of this simple example.

In the previous code examples, the text size picker is in the same component where the custom property is declared. In a real application, we would use a service to communicate the state change to the root component from any component that contains the text size picker.

The text size picker and its effect on a paragraph's font size can be seen in the following screenshot:

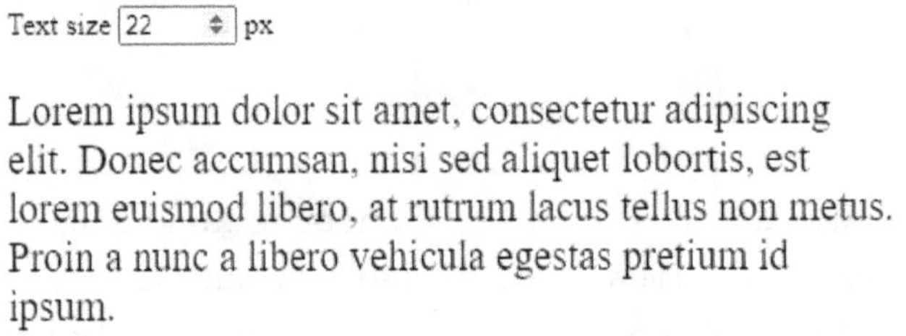

Text size 22 px

Lorem ipsum dolor sit amet, consectetur adipiscing elit. Donec accumsan, nisi sed aliquet lobortis, est lorem euismod libero, at rutrum lacus tellus non metus. Proin a nunc a libero vehicula egestas pretium id ipsum.

Figure 3.4 – Text size picker

This section started with a simple example that demonstrated CSS custom property scopes through background colors. In the second example, we saw how they can be repurposed for non-styling concerns, such as globalization.

Finally, we saw how we can change the value of a CSS custom property at runtime; for example, to select the text size of an application. We are going to apply this knowledge to implement a feature in the application that will be used in *Part 2 , Build a Real-World Application with the Angular Ivy Features You Learned.*

In the next section, we explore the provider scope shorthands introduced by Angular Ivy. These will also prove useful for the features we are going to implement in *Part 2 , Build a Real-World Application with the Angular Ivy Features You Learned.*

Discovering the any and platform provider scopes

Tree-shakable providers make it possible to create dependencies that are removed from a compiled bundle if they have not been unused. This is especially important for Angular libraries, but it also plays a role in large Angular applications and **monorepositories** that contain multiple Angular applications.

As an experienced Angular developer, you are probably already familiar with the `root` provider scope shorthand, which can be passed to the `providedIn` option in the `Injectable` decorator factory, and the `InjectionToken` constructor. It is used to create an application-wide singleton dependency.

Angular version 9 introduced two new provider scope shorthands, namely `any` and `platform`. In this section, we are going to discuss `any`, which is the most complex of the two provider scope shorthands.

The any provider scope

The any provider scope declares a singleton dependency per module injector. This means that the root module – typically, `AppModule` – and all Angular modules that it eagerly loads through static import statements will share a single instance.

Any lazy-loaded Angular feature module and all the Angular modules it statically imports will share a second dependency instance. A different Angular feature module and its static Angular module dependencies will share a third dependency instance, and so on.

> **Important Note**
>
> Like the `root` shorthand, dependencies provided using the `any` provider scope shorthand are tree-shakable. This separates it from Angular service module providers, Angular feature module providers, and the `forRoot-forChild` pattern. In the alternative patterns, the dependency is always bundled, even if unused, because a static reference between an Angular module and the dependency exists.

If we think carefully about the `any` provider scope, we will realize that it is for very specific use cases.

The following are the traits that describe the `any` provider scope shorthand:

- It is useful for controlling stateful dependencies that vary between lazy-loaded Angular feature modules.
- It is useful for analytics, configurations, logging, and metrics.
- It is a tree-shakable alternative of the following:
  - The `forRoot-forChild` pattern
  - A lazy-loaded Angular feature module provider
  - An Angular service module provider
- It keeps the dependency tree-shakable in contrast to the alternatives, as described in the previous *Important Note*.

Let's consider one of these use cases. Say we have a backend configuration that can either vary between lazy-loaded Angular features or fall back to a common configuration for lazy-loaded Angular features that have no backend configuration:

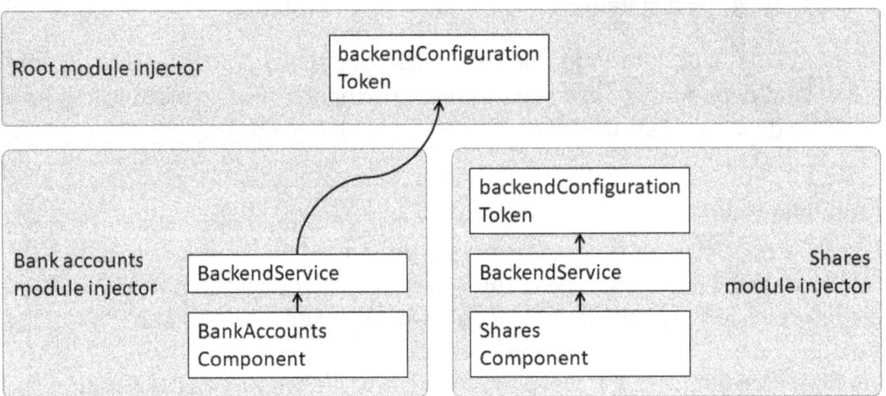

Figure 3.5 – A dependency hierarchy that uses the any provider scope for the backend service

The preceding diagram illustrates an example in which the backend service uses the any provider scope. The bank accounts module injector creates an instance of `BackendService`, which is passed to `BankAccountsComponent`. This feature does not provide a separate backend configuration, so it falls back to the one provided by the root module injector.

The shares Angular feature module provides a `BackendConfiguration` value for `backendConfigurationToken`, which is passed to the `BackendService` instance that the bank accounts module injector creates for `SharesComponent`.

Let's look at the most interesting parts of the implementation for this example. The full implementation can be found in this book's companion GitHub repository:

1. First, we define the shape of the backend configuration:

```
export interface BackendConfiguration {
  readonly baseUrl: string;
  readonly retryAttempts: number;
  readonly retryIntervalMs: number;
}
```

2. Next, we create a backend service that wraps Angular's `HttpClient` service, prefixes the URL with the configured base URL, and adds a retry strategy based on the backend configuration. Note how it uses the `any` provider scope shorthand to do this:

```
import { HttpClient } from '@angular/common/http';
import { Inject, Injectable } from '@angular/core';
import { Observable } from 'rxjs';

import { BackendConfiguration } from './backend-
configuration';
import { backendConfigurationToken } from './backend-
configuration.token';
import { BackendGetOptions } from './backend-get-
options';
import { retryWithDelay } from './retry-with-delay.
operator';

@Injectable({
  providedIn: 'any',
})
export class BackendService {
  constructor(
    @Inject(backendConfigurationToken)
    private configuration: BackendConfiguration,
    private http: HttpClient
  ) {}

  get<T>(url: string, options: BackendGetOptions =
   {}): Observable<T> {
    return this.http
      .get<T>(`${this.configuration.baseUrl}/${url}`,
      {
        ...options,
        responseType: 'json',
```

```
      })
      .pipe(
        retryWithDelay(
          this.configuration.retryAttempts,
          this.configuration.retryIntervalMs
        )
      );
  }
}
```

3. For reference, this is our `retryWithDelay` operator:

```
import { concat, throwError } from 'rxjs';
import { delay, retryWhen, take } from 'rxjs/operators';

export const retryWithDelay = <T>(retries: number,
delayMs: number) =>
  retryWhen<T>((errors) =>
    concat(
      errors.pipe(delay(delayMs), take(retries)),
      throwError(`Failed after ${retries} retries.`)
    )
  );
```

4. Our root Angular module, `AppModule`, lists its imports first:

```
import { HttpClientModule } from '@angular/common/http';
import { NgModule } from '@angular/core';
import { BrowserModule } from '@angular/platform-
browser';

import { AppComponent } from './app.component';
import { BackendConfiguration } from './data-access/
backend-configuration';
import { backendConfigurationToken } from './data-access/
backend-configuration.token';
import { AppRoutingModule } from './app-routing.module';
```

5. Then, it lists its backend configuration, which will be shared with any lazy-loaded Angular feature that does not provide its own backend configuration:

```
const backendConfiguration: BackendConfiguration = {
  baseUrl: 'https://api01.example.com',
  retryAttempts: 4,
  retryIntervalMs: 250,
};
```

6. `AppModule` declares `AppComponent` and provides common singleton dependencies through Angular modules listed in its imports option array. Additionally, it imports `AppRoutingModule` before it provides its backend configuration for `backendConfigurationToken`:

```
@NgModule({
  bootstrap: [AppComponent],
  declarations: [AppComponent],
  imports: [BrowserModule, HttpClientModule,
   AppRoutingModule],
  providers: [
    { provide: backendConfigurationToken, useValue:
      backendConfiguration },
  ],
})
export class AppModule {}
```

7. For reference, this is the backend configuration token:

```
import { InjectionToken } from '@angular/core';

import { BackendConfiguration } from './backend-
configuration';

export const backendConfigurationToken = new
InjectionToken<
  BackendConfiguration
>('Backend configuration');
```

8. `SharesModule` provides a backend configuration that points to a different API URL than the one configured in `AppModule`. `SharesModule` also configures a shorter retry interval and a higher number of retry attempts. This makes sense as shares properties change more frequently than other parts of a banking application:

```
import { CommonModule } from '@angular/common';
import { NgModule } from '@angular/core';
import { RouterModule, Routes } from '@angular/router';

import { BackendConfiguration } from '../data-access/
backend-configuration';
import { backendConfigurationToken } from '../data-
access/backend-configuration.token';
import { SharesComponent } from './shares.component';

const routes: Routes = [{ path: '', component:
SharesComponent }];
const backendConfiguration: BackendConfiguration = {
  baseUrl: 'https://api02.example.com',
  retryAttempts: 7,
  retryIntervalMs: 100,
};

@NgModule({
  declarations: [SharesComponent],
  imports: [CommonModule,
   RouterModule.forChild(routes)],
  providers: [
    { provide: backendConfigurationToken, useValue:
     backendConfiguration },
  ],
})
export class SharesModule {}
```

9. `SharesComponent` injects `BackendService`, which it uses to query for shares. The backend service instance is scoped to the `SharesModule` module injector. Because a new instance has been created, when it injects `BackendConfiguration` using `backendConfigurationToken`, it resolves to the configuration provided in `SharesModule`, as we saw in the previous step:

```
import { Component } from '@angular/core';

import { BackendService } from '../data-access/backend.
service';

@Component({
  selector: 'workspace-shares',
  templateUrl: './shares.component.html',
})
export class SharesComponent {
  shares$ = this.backend.get('shares');

  constructor(private backend: BackendService) {}
}
```

10. `BankAccountsModule` is very similar to `SharesModule`, but it does not provide a backend configuration:

```
import { CommonModule } from '@angular/common';
import { NgModule } from '@angular/core';
import { RouterModule, Routes } from '@angular/router';

import { BankAccountsComponent } from './bank-accounts.
component';

const routes: Routes = [{ path: '', component:
BankAccountsComponent }];

```

```
@NgModule({
  declarations: [BankAccountsComponent],
  imports: [RouterModule.forChild(routes),
   CommonModule],
})
export class BankAccountsModule {}
```

11. `BankAccountsComponent` is almost identical to `SharesComponent`. It injects an instance of `BackendService` that it uses to query for bank accounts. Remember that `BackendService` has the any provider scope, so this instance is scoped to `BankAccountsModule`. However, it resolves the backend configuration provided by `AppModule` because a backend configuration is not provided in the bank accounts Angular module:

```
import { Component } from '@angular/core';

import { BackendService } from '../data-access/backend.
service';

@Component({
  selector: 'workspace-bank-accounts',
  templateUrl: './bank-accounts.component.html',
})
export class BankAccountsComponent {
  bankAccounts$ = this.backend.get('bank-accounts');

  constructor(private backend: BackendService) {}
}
```

This is a lot to keep track of, so we will show the dependency hierarchy diagram from earlier again here:

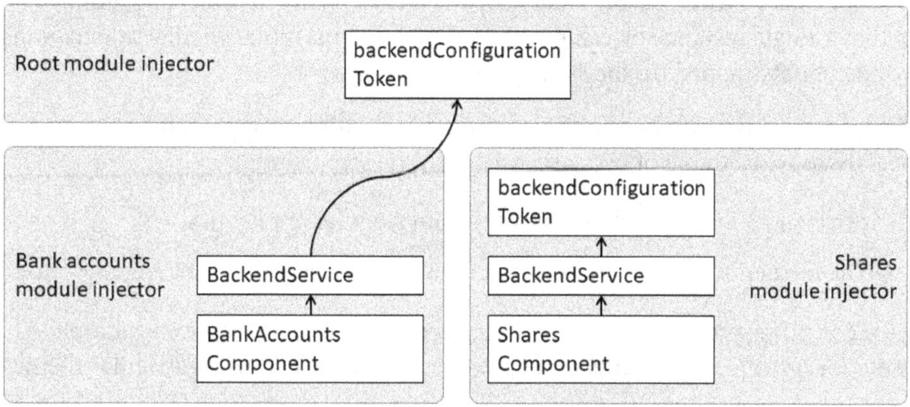

Figure 3.6 – A dependency hierarchy that uses the any provider scope for the backend service

Two backend configurations are registered through the same `backendConfigurationToken` – one in the root Angular module injector and one in the shares Angular module injector.

There are two `BackendService` instances – one that is scoped to the bank accounts Angular module injector and one that is scoped to the shares Angular module injector.

The backend service that's used by `BankAccountsComponent` uses the backend configuration, which depends on the backend configuration provided in the root Angular module injector. The backend service that's used by `SharesComponent` depends on the backend configuration provided in the shares Angular module injector.

Hopefully, this makes sense to you now. If not, go back and read the step-by-step instructions again while consulting the dependency hierarchy or browse through the `any` provider scope application in this book's companion GitHub repository, as mentioned in the introduction to this chapter.

With an understanding of the `any` provider scope, we will move on to the second provider scope introduced by Angular Ivy: the `platform` provider scope.

The platform provider scope

The `platform` provider scope shorthand is conceptually less complex than the `any` provider scope. The `platform` provider scope creates a platform-wide singleton, meaning that a single instance or value is shared between multiple Angular applications that have been bootstrapped on the same page.

While it might not sound very common to bootstrap multiple Angular components on the same page, there are a couple of use cases where this is important:

- Microfrontends, where each one is a separate Angular application
- Angular Elements, where each web component has a separate root Angular module

Using the `platform` provider scope, we can share a dependency between multiple microfrontends or web components that have been implemented using Angular Elements.

To visualize how dependencies flow when using the `platform` provider scope, imagine that we have two Angular applications that are bootstrapped on the same page. Let's say we have a documents application and a music application:

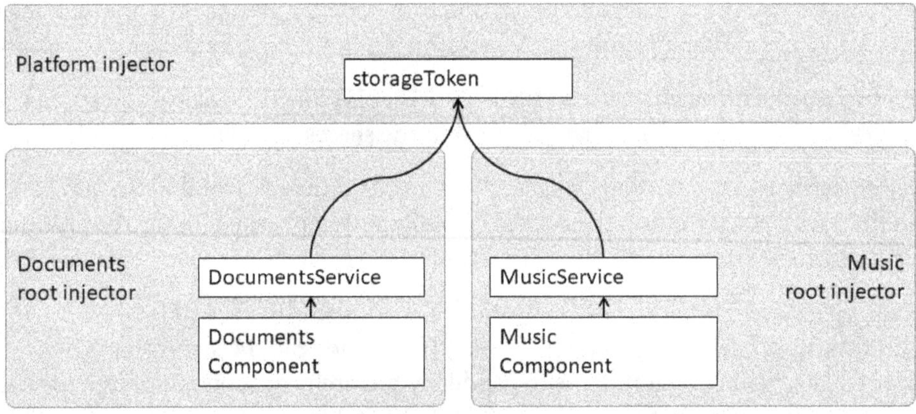

Figure 3.7 – A dependency hierarchy that uses the platform provider scope for the Storage API

In the preceding diagram, we can see how both `DocumentsService` and `MusicService` depend on `storageToken`, even though they are provided in separate applications:

```
import { InjectionToken } from '@angular/core';

export const storageToken =
  new InjectionToken<Storage>('Web Storage API', {
  factory: (): Storage => localStorage,
    providedIn: 'platform',
  });
```

`storageToken` represents the `Storage` API. Because this API is shared between all the applications on the same page, we provide it using the `platform` provider scope shorthand, as seen in the preceding code.

As indicated by the following code, provider shorthands work for both class-based services and dependency injection tokens:

```
import { Inject, Injectable } from '@angular/core';

import { storageToken } from './storage.token';

@Injectable({
  providedIn: 'root',
})
export class DocumentsService {
  constructor(
     @Inject(storageToken) private storage: Storage,
  ) {}

  loadDocument(id: number) {
    return this.storage.getItem(`documents/${id}`);
  }
}
```

In the preceding code, `DocumentsService` is an application-wide singleton service, which means that it is scoped to the documents root injector. The documents root injector is scoped to the documents application, which is bootstrapped separately from the music application.

Our classes do not need to do anything special to depend on a platform-scoped dependency. For example, in the previous code, we injected `storageToken` as usual. However, depending on the use case, the storage token reference is shared through the global scope or a shared bundle.

This simple example demonstrated the `platform` provider scope and shorthand, which concludes this section.

Summary

This chapter prepared you for implementing features in *Part 2, Build a Real-World Application with the Angular Ivy Features You Learned* by discussing CSS Custom Properties in detail, as well as the `any` and `platform` provider scopes.

First, we discussed how CSS Custom Properties are native, scoped runtime CSS variables that do not rely on specificity or CSS source ordering. Through a simple example, we demonstrated how variables and element trees can be combined for stylistic purposes.

CSS Custom Properties can hold any value. For example, we explored an example of using them for globalization and another example for controlling the text size dynamically at runtime.

The next topic we discussed was the `any` provider scope. This scope defines a boundary around each module injector. A dependency instance or value is created per application chunk and shared throughout its module injector.

We learned that the `any` provider scope is useful for orchestrating stateful dependencies, maybe for analytics, configurations, logging, and metrics. The `any` provider scope is a tree-shakable alternative to other providers such as the `forRoot-forChild` pattern, a lazy-loaded Angular feature module provider, or an Angular service module provider.

To demonstrate this, we created a backend configuration that could be specified per feature module injector, as well as the root module injector since each module injector gets its own instance of `BackendService`.

Finally, we discussed the `platform` provider scope, which is used to share a common dependency such as a native API between multiple Angular applications or web components implemented using Angular Elements.

We took a brief look at an imaginary example of a documents application and a music application running in parallel on the same site or maybe one at a time. They both share the same dependency value represented by the `storageToken` dependency injection token. The lifetime of this dependency outlives that of both applications.

In the next chapter, we are going to explore the major Angular Components features introduced by Ivy, including the official Google product components, the Clipboard API, and component testing harnesses. We are going to use all of them in *Part 2, Build a Real-World Application with the Angular Ivy Features You Learned.*

4
Exploring Angular Components Features

In *Chapter 3, Introducing CSS Custom Properties and New Provider Scopes*, we explored new platform and framework features in depth to prepare for *Part 2, Build a Real-World Application with the Angular Ivy Features You Learned*. Let's continue, but this time focus on brand new APIs introduced in the Angular component packages.

With Angular Ivy, the two first official Angular packages for Google products were introduced: YouTube Player and embedded Google Maps. We will explore both of these packages in this chapter.

Finally, we will cover two new APIs introduced by the Angular CDK: the Clipboard API and component testing harnesses. The Angular CDK clipboard directive, service, and domain object interact with the operating system's native clipboard. A component harness is a testing API wrapping one or more Angular components using the test-as-a-user approach. It is usable in the context of unit tests and end-to-end tests.

This chapter covers the following topics:

- Angular YouTube Player
- Angular Google Maps components
- The Clipboard API
- Testing as a user with component harnesses

Learning about these topics will enable you to use these powerful features for our existing application in *Part 2, Build a Real-World Application with the Angular Ivy Features You Learned.*

Technical requirements

To proceed with this chapter, you will need the following:

- Angular 9.0
- TypeScript 3.6

You can find the complete code for the video and map examples in this book's companion GitHub repository at `https://github.com/PacktPublishing/Accelerating-Angular-Development-with-Ivy/tree/main/projects/chapter4`.

Angular YouTube Player

As part of Angular Ivy, the Angular team published official Angular components for Google products. One of them is the Angular YouTube Player. As the name suggests, it is used to embed a YouTube video player in your Angular application while getting the convenience of Angular data binding as well as programmatic access to the YouTube Player API through a component reference.

In this section, we will go through the setup needed to start using the Angular YouTube Player. We will then look at its entire API to get familiar with its capabilities and usage.

Getting started

First, make sure to install the `@angular/youtube-player` package by using the following command:

```
ng add @angular/youtube-player
```

Now add `YouTubePlayerModule` to the module that declares the component that is going to use the YouTube Player component, as shown in the following code:

```
import { NgModule } from '@angular/core';
import { YouTubePlayerModule } from '@angular/youtube-player';

import { VideoComponent } from './video.component';

@NgModule({
  declarations: [VideoComponent],
  exports: [VideoComponent],
  imports: [YouTubePlayerModule],
})
export class VideoModule {}
```

Before using the Angular YouTube Player component for the first time, we must load the YouTube IFrame API script. It is a 100 KB script, so depending on your priorities, there are several ways to load it. You could load it as part of the `scripts` option in your application project configuration in `angular.json`, but then the user would always pay the price upfront for loading, parsing, and executing this script.

Instead, we can load it in when the component using it is activated, as we will see in the following tutorial:

1. First, we import the Angular core and common APIs we need and declare component metadata and the component name, in this case, `VideoComponent`:

```
import { DOCUMENT } from '@angular/common';
import { Component, Inject, OnDestroy, OnInit } from '@
angular/core';

@Component({
  selector: 'app-video',
  templateUrl: './video.component.html',
})
export class VideoComponent implements OnDestroy, OnInit
{
```

2. Next, we add a private property to keep a reference to the script we are going to dynamically load:

```
#youtubeIframeScript: HTMLScriptElement;
```

3. Now, we inject the `Document` object of our platform, that is, the browser or the server:

```
constructor(@Inject(DOCUMENT) private document:
 Document) {
```

4. We create an `async` script element that points to the `https://www.youtube.com/iframe_api` URL. This is a loader script for the initialization script, which sets up the YouTube API needed to embed YouTube videos. The Angular YouTube Player component is a convenient wrapper around this API:

```
  this.#youtubeIframeScript =
   this.document.createElement('script');
  this.#youtubeIframeScript.src =
   'https://www.youtube.com/iframe_api';
  this.#youtubeIframeScript.async = true;
}
```

5. When `VideoComponent` is initializing, add the YouTube IFrame API script element to the body element of the HTML document:

```
ngOnInit(): void {
  this.document.body.appendChild(
   this.#youtubeIframeScript);
}
```

6. We remove the script element when `VideoComponent` is deactivated:

```
  ngOnDestroy(): void {
    this.document.body.removeChild(
     this.#youtubeIframeScript);
  }
}
```

Do not worry about loading the script multiple times. The browser will cache it and the YouTube IFrame API script will detect that it has already been loaded. For an even more robust setup solution, we can listen for the `loaded` event on the script element, set a loaded flag in our code, and add a condition not to load the script the next time this component is activated.

This is all the setup we need to use the Angular YouTube Player component. The full example is available in this book's companion GitHub repository, as mentioned in the introduction of this chapter. Now, let's move on to usage instructions.

Usage

The simplest example of using the Angular YouTube Player is to put the following code snippet in a component template:

```
<youtube-player videoId="8NQCgmAQEdE"></youtube-player>
```

We pass a YouTube video ID to the `videoId` input property and the Angular YouTube Player will take care of everything else. Of course, it also allows for more customization. Let's start by looking at the `YouTubePlayer` component's data binding API.

API reference

As of Angular version 12, the YouTube Player has no API reference. This is provided for you in this section so that you do not have to look up the source code and cross-reference it with the online YouTube JavaScript API reference to use it.

The data binding API

With the data binding API, we can declaratively configure the Angular YouTube Player component. The input properties are used to configure settings while the output properties emit events about user interactions and the video environment.

First, we will look at the input properties.

Input properties

The input properties are used to configure the playback and visuals of the embedded player:

- `@Input() endSeconds: number | undefined;`

  To set a playback ending point of the YouTube video, pass the offset in the number of seconds from the beginning of the video.

- `@Input() height: number | undefined;`

  The height of the YouTube Player is specified using the `height` input property, in the number of CSS pixels. It defaults to `390`.

- `@Input() playerVars: YT.PlayerVars | undefined;`

  Many additional options can be passed to the `playerVars` input property, for example, we can pass `{ modestbranding: YT.ModestBranding.Modest }` to hide the YouTube logo. For a full reference, see `https://developers.google.com/youtube/player_parameters.html?playerVersion=HTML5#Parameters`.

- `@Input() showBeforeIframeApiLoads: boolean | undefined;`

  The `showBeforeIframeApiLoads` input property defaults to `false` but can be set to `true` to make the YouTube Player component throw an error if it is loaded before the YouTube IFrame API is loaded.

- `@Input() startSeconds: number | undefined;`

  To set a playback starting point for the YouTube video, pass the offset in the number of seconds from the beginning of the video.

- `@Input() suggestedQuality: YT.SuggestedVideoQuality | undefined;`

  The `suggestedQuality` input property accepts one of the following quality identifiers: `'default'`, `'small'`, `'medium'`, `'large'`, `'hd720'`, `'hd1080'`, `'highres'`.

- `@Input() videoId: string | undefined;`

  The `videoId` input property accepts the YouTube video ID of the video to be played back.

- `@Input() width: number | undefined;`

  The width of the YouTube Player is specified using the `width` input property, in the number of CSS pixels. It defaults to `640`.

Next, we will look at the output properties, which emit events about the user interactions and video environment.

Output properties

The output properties expose the events that are emitted by the YouTube IFrame API. For a full reference of the events, see `https://developers.google.com/youtube/iframe_api_reference#Events`:

- `@Output() apiChange: Observable<YT.PlayerEvent>;`

  When the closed captioning module is loaded or unloaded, an event is emitted by the `apiChange` output property.

- `@Output() error: Observable<YT.OnErrorEvent>;`

  The `error` output property emits an event when one of the following errors occurs, all of which are accessible through the `YT.PlayerError` enum: `EmbeddingNotAllowed`, `EmbeddingNotAllowed2`, `Html5Error`, `InvalidParam`, `VideoNotFound`.

- `@Output() playbackQualityChange: Observable<YT.OnPlaybackQualityChangeEvent>;`

  When the playback quality changes, one of the following quality identifiers is emitted through the `playbackQualityChange` output property: `'default'`, `'small'`, `'medium'`, `'large'`, `'hd720'`, `'hd1080'`, `'highres'`.

- `@Output() playbackRateChange:Observable<YT.OnPlaybackRateChangeEvent>;`

  When the video playback rate is changed, the `playbackRateChange` output property emits an event where its `data` property is a number such as `1.0` signifying the playback speed.

- `@Output() ready: Observable<YT.PlayerEvent>;`

  This event is emitted when the YouTube Player is fully loaded and ready to be controlled.

- `@Output() stateChange: Observable<YT.OnStateChangeEvent>;`

  An event is output every time the YouTube Player state changes to one of the following, which are all accessible through the `YT.PlayerState` enum: `BUFFERING`, `CUED`, `ENDED`, `PAUSED`, `PLAYING`, `UNSTARTED`.

Our components can send commands or read information from the YouTube Player using its components methods, which we will discuss next.

Component methods

The `YouTubePlayer` component has several public methods that we can use to control the embedded video player:

- `getAvailablePlaybackRates(): number[];`

  Determines the supported playback rate for the current active or queued video. For a video supporting only normal playback speed, `[1.0]` is returned. Something like `[0.25, 0.5, 1.0, 1.5, 2.0]` might be returned.

- `getCurrentTime(): number;`

  Determines the number of seconds elapsed since the beginning of the video.

- `getDuration(): number;`

  Determines the duration in seconds of the current active video. If the video's metadata has not been loaded, it will return `0`.

  For live streams, it will return the number of seconds since the stream started, was reset, or interrupted.

- `getPlaybackRate(): number;`

  Determines the playback rate of the current active or queued video where `1.0` is the normal playback speed.

- `getPlayerState(): YT.PlayerState | undefined;`

  Determines the current player state between one of the following, which are all accessible through the `YT.PlayerState` enum: `BUFFERING`, `CUED`, `ENDED`, `PAUSED`, `PLAYING`, `UNSTARTED`. Returns the latest value represented by an event emitted by the `stateChange` output property.

- `getVideoEmbedCode(): string;`

  Determines the HTML markup needed to embed the video in an HTML page, for example:

```
<iframe id="ytplayer" type="text/html" width="720"
height="405"
src="https://www.youtube.com/embed/8NQCgmAQEdE"
frameborder="0" allowfullscreen></iframe>
```

- `getVideoLoadedFraction(): number;`

  Determines the percentage of the video that has been buffered by the player where `0.0` is 0% and `1.0` is 100%.

- `getVideoUrl(): string;`

  Determines the video's full URL on `youtube.com`.

- `getVolume(): number;`

  Determines the volume level between `0` and `100`. Only returns integers. When muted, this method will return the level that was active when the audio was muted.

- `isMuted(): boolean;`

  Checks whether the audio is muted. `true` means muted, `false` means unmuted.

- `mute(): void;`

  Mutes the audio.

- `pauseVideo(): void;`

  Pauses the video. An event is emitted through `stateChange`.

- `playVideo(): void;`

  Starts playing the video. Does not count towards the video's view count on YouTube. An event is emitted through `stateChange`.

- `seekTo(seconds: number, allowSeekAhead: boolean): void;`

  Goes to the specified timestamp. The video will keep being paused if paused before seeking. Setting `allowSeekAhead` to `false` keeps the player from downloading unbuffered content from the server. This can be used in combination with a progress bar.

- `setPlaybackRate(playbackRate: number): void;`

  Adjusts the playback speed. Only affects the current active or queued video. Passing `1.0` to `playbackRate` sets the playback speed to normal.

  We should first call `getAvailablePlaybackRates` to check which playback rates are supported for the video. Listen to events emitted by the `playbackRateChange` output property to verify that the playback speed was successfully adjusted.

  If the passed `playbackRate` does not exactly match supported playback speeds, the nearest rate will be matched, rounding to `1.0`.

- `setVolume(volume: number): void;`

  Adjusts the volume to a level between `0` and `100`. Only accepts integers.

- `stopVideo(): void;`

  Stops loading or playing the video. We can use this if we know that the user will not be watching additional videos in the YouTube Player. It's not necessary to call it before playing a different video. An event is emitted through `stateChange`, but the state could be any of `CUED`, `ENDED`, `PAUSED`, or `UNSTARTED`.

- `unMute(): void;`

  Unmutes the audio.

With the knowledge of the full component API of the `YouTubePlayer` component, you can build your own controls on top of it, configure default settings across all YouTube Players in our application, control many YouTube Players at the same time or implement a YouTube video snippet widget using Angular.

We have discussed how to install and set up your application for the Angular YouTube Player. We have seen a simple example usage, listed its full API, and discussed its use cases. You are now ready to use the Angular YouTube Player in *Part 2, Build a Real-World Application with the Angular Ivy Features You Learned.*

Next, we will look at the official Angular components for Google Maps.

Angular Google Maps components

In this section, we will look at the official Google product component package called Angular Google Maps. The Google Maps API is large, so this package includes both a `GoogleMap` component and several other components used to configure and control its many features.

It consists of the following Angular components and directives:

- `GoogleMap`
- `MapBicyclingLayer`
- `MapCircle`
- `MapGroundOverlay`
- `MapInfoWindow`
- `MapKmlLayer`
- `MapMarker`
- `MapMarkerClusterer`
- `MapPolygon`
- `MapRectangle`
- `MapTrafficLayer`
- `MapTransitLayer`

We will explore the necessary component, `GoogleMap`, and the commonly used components `MapInfoWindow`, `MapMarker`, and `MapMarkerClusterer`.

Getting started

To use the Angular Google Maps components, we first must load the Google Maps JavaScript API. This example wrapper component illustrates how to conditionally render the Google Maps component after the Google Maps JavaScript API has been initialized:

```
import { Component, ViewChild } from '@angular/core';
import { HttpClient } from '@angular/common/http';
import { GoogleMap }from '@angular/google-maps';
import { Observable, of } from 'rxjs';
import { catchError, mapTo } from 'rxjs/operators';
import { AppConfig } from '../app-config';

```

```
@Component({
  selector: 'app-map',
  templateUrl: './map.component.html',
})
export class MapComponent {
  @ViewChild(GoogleMap, { static: false })
map?: GoogleMap;

isGoogleMapsApiLoaded$: Observable<boolean> = this.http.
jsonp('https://maps.googleapis.com/maps/api/js?key=${this.
config.googleMapsApiKey}','callback').pipe(
mapTo(true),
   catchError(() => of(false)),
  );

  constructor(
    private config: AppConfig,
    private http: HttpClient,
  ) {}
}
```

Our `MapComponent` has an observable UI property called `isGoogleMapsApiLoaded$`, which loads the Google Maps JavaScript API with a pre-configured API key. We use this to conditionally render the `GoogleMap` component in the component template, as shown in the following code:

```
<google-map *ngIf="isGoogleMapsApiLoaded$ | async; else
spinner"></google-map>

<ng-template #spinner>
<mat-spinner></mat-spinner>
</ng-template>
```

Until the Google Maps JavaScript API is loaded, an Angular Material Spinner component is shown.

Notice that we created a view child query for `GoogleMap` and stored it in the `map` property. This can be used to programmatically control the map from the component model.

For reference, here is the Angular module that declares our example `MapComponent`:

```
import { CommonModule } from '@angular/common';
import { HttpClientModule, HttpClientJsonpModule } from '@
angular/common/http';
import { NgModule } from '@angular/core';
import { GoogleMapsModule } from '@angular/google-maps';
import { MatProgressSpinnerModule } from'@angular/material/
progress-spinner';
import { MapComponent } from './map.component';

@NgModule({
  declarations: [MapComponent],
  exports: [MapComponent],
  imports: [
    HttpClientModule,
    HttpClientJsonpModule,
    GoogleMapsModule,
    MatProgressSpinnerModule,
  ],
})
export class MapModule {}
```

Now that we have all the setup needed to work with the Angular Google Maps API, let's take a closer look at the most common components included in the `@angular/google-maps` package.

The GoogleMap component

The `GoogleMap` component is the primary entry point to the Angular Google Maps package. It is the top-level component that can contain other components from this package.

This component is a declarative, Angular-specific wrapper for the `google.maps.Map` class from the Google Maps JavaScript API. Refer to the API reference for more details on the `Map` class (`https://developers.google.com/maps/documentation/javascript/reference/map`).

The Google map component has the center, height, mapTypeId, width, and zoom input properties. It also accepts an options input of the google.maps.MapOptions type. It has 19 different output properties, all matching DOM events described in the Google Maps JavaScript API reference for the Map class. It also has a wide range of methods available for controlling the map.

The MapMarker component

The MapMarker component's element, <map-marker>, is either nested inside the <google-map> element or a <map-marker-clusterer> element.

This component represents a marker on a Google map. We can use a label, a marker icon, or a marker symbol to customize it.

MapMarker is an Angular-specific wrapper for the google.maps.Marker class. It has the clickable, label, position, and title input properties. It also accepts an options input of the google.maps.MarkerOptions type.

We can pass a custom icon through the marker options, for example, as follows where we use a beach flag icon:

```
<google-map
  [center]="{ lat: 56.783778, lng: 8.228937 }"
>
  <map-marker
    [options]="{ icon: 'https://developers.google.com/
     maps/documentation/javascript/examples/full/
     images/beachflag.png' }"
    [position]="{ lat: 56.783778, lng: 8.228937 }"
  ></map-marker>
</google-map>
```

Refer to the API reference for more details on the Marker class (https://developers.google.com/maps/documentation/javascript/reference/marker).

The MapMarkerClusterer component

The `MapMarkerClusterer` component is an Angular-specific wrapper around the `MarkerClusterer` class from the `@googlemaps/markerclustererplus` package. Its element, `<map-marker-clusterer>`, is nested inside of `<google-map>` and contains multiple `<map-marker>` elements.

This component is used to group many map markers into clusters when zoomed out on a map.

Before we can use it, we must load it into a global variable by inserting the following script tag at an appropriate place, for simplicity's sake in the template of the component using it:

```
<script src="https://unpkg.com/@googlemaps/markerclustererplus/
dist/index.min.js"></script>
```

The `MapMarkerClusterer` component has 18 different input properties, such as `minimumClusterSize`, `maxZoom`, and `zoomOnClick`. The `imagePath` input property can be used to specify custom map marker cluster images where this path is automatically suffixed with `[1-5].png` by default.

Two output properties are available: `clusteringbegin` and `clusteringend`. They are emitted whenever markers first cluster and when they are split into individual markers, respectively.

Refer to the API reference for more details on the `MarkerClusterer` class (`https://developers.google.com/maps/documentation/javascript/marker-clustering`).

The MapInfoWindow component

The `MapInfoWindow` component is an Angular-specific wrapper for the `google.maps.InfoWindow` class. Its element, `<map-info-window>`, is nested inside of `<google-map>`.

It is an overlay used to display notifications or metadata on top of a map, usually near a map marker.

Its `position` input property declares where on the map it appears. Additionally, it accepts an `options` input of type `google.maps.InfoWindowOptions`. It has five different output properties – `closeClick`, `contentChanged`, `domready`, `positionChanged`, and `zindexChanged` – all matching DOM events described in the Google Maps JavaScript API reference for the `InfoWindow` class.

The `MapInfoWindow` component uses content projection, which means that the content we put inside of its custom element tags is rendered in its overlay when opened.

To display a `MapInfoWindow` component, call its `open` method, which optionally accepts `MapMarker` that the info window will be attached to. The `close` method hides the `MapInfoWindow` component.

Refer to the API reference for more details on the `InfoWindow` class (`https://developers.google.com/maps/documentation/javascript/reference/info-window`).

Now that you have an overview of the most used parts of the official Google Maps Angular components, you are prepared to use Google Maps in the Angular application in *Part 2, Build a Real-World Application with the Angular Ivy Features You Learned*.

In the next section, we will learn about the Angular CDK's Clipboard API.

The Clipboard API

The Angular CDK's Clipboard API offers a directive and a service to interact with the operating system's clipboard through the browser. The `CdkCopyToClipboard` directive can be used declaratively while the `Clipboard` service is used for use cases where a programmatic API is a better fit. The Clipboard API additionally takes care of long texts through the `PendingCopy` class.

In this section, you will learn how to use each of these classes from the Angular CDK package.

The CdkCopyToClipboard directive

The `CdkCopyToClipboard` directive is exported by `ClipboardModule`, which is in the `@angular/cdk/clipboard` sub-package. Its directive selector is `[cdkCopyToClipboard]`. The directive has an input property of the same name as the directive, which accepts the text that is copied when the element it is attached to is clicked.

Because of browser security concerns, copying text to the clipboard must be done following a click event triggered by a user.

The copy to clipboard directive additionally has an input property named `cdkCopyToClipboardAttempts`. It accepts a number, which is the number of macrotask cycles the directive will attempt to copy the text for before giving up. This is relevant in the case of bigger text because of an implementation detail that ensures cross-browser compatibility until the upcoming Clipboard API is supported across all major browsers. We will discuss this caveat further when exploring the `PendingCopy` class.

The copy to clipboard directive and its retry parameter is demonstrated in the following code snippet:

```
<button
  [cdkCopyToClipboard]="transactionLog"
  [cdkCopyToClipboardAttempts]="5"
>
  Copy transaction log
</button>
```

Finally, the `CdkCopyToClipboard` directive has an output property named `cdkCopyToClipboardCopied`, which emits a Boolean value every time copying to the clipboard is attempted and indicates whether copying succeeded.

The Clipboard service

The `Clipboard` service is useful when we want to perform other operations before or after copying text to the clipboard, if the text is not easily accessible from a component template, or if we want more fine-grained control when copying big texts.

The clipboard service has two methods. The `Clipboard#copy` method accepts the text to copy to the clipboard and returns a Boolean value indicating whether the copy operation was successful.

For some large texts, the `Clipboard#copy` method fails and we have to use the `Clipboard#beginCopy` method instead. This method also accepts the text that we want to copy to the clipboard but returns an instance of the `PendingCopy` class that we must interact further with to follow through on the copy to clipboard operation. This class is discussed next.

The PendingCopy class

An instance of the `PendingCopy` class is returned from the `Clipboard#beginCopy` method. As mentioned earlier in this section, this has to do with implementation details ensuring cross-browser compatibility for copying large texts.

The first thing we must learn about the `PendingCopy` class is that we must tear down all instances by calling the `PendingCopy#destroy` method once we have finished using them or our application will leak resources.

The `PendingCopy#copy` method accepts no arguments and returns a Boolean value indicating whether the copy to clipboard operation succeeded. If `false` is returned, we should schedule another attempt for later.

As described earlier in this section, the `CdkCopyToClipboard` directive supports a retry strategy for copying large texts by passing a maximum number of attempts to its `cdkCopyToClipboardAttempts` input property.

Now that we have discussed all parts of the Angular CDK's Clipboard API, we are ready to implement a feature for the hands-on application in *Part 2, Build a Real-World Application with the Angular Ivy Features You Learned.*

In the next section, you will learn about component testing harnesses, an innovative API for testing components, and authoring testing APIs for components exposed in library packages.

Testing as a user with component harnesses

The Angular CDK's API for authoring and using component testing harnesses is a fresh approach with the test-as-a-user philosophy in mind. Each component or related set of components can have a component harness for tests. A component harness is a testing API for interacting with those components that can be used in unit, integration, and end-to-end tests.

Component testing harnesses internally rely only on a single selector for the component they wrap. Library authors can publish component harnesses for their Angular components. In this way, their consumers' tests, which depend on the library's components, will not have dependencies on the DOM structure except for that one selector, which the library authors are able to change if needed.

This is exactly what the Angular Components team do for Angular CDK and Angular Material. They release and maintain component harnesses for all their Angular components.

Harness environments and harness loaders

A harness environment represents the context tests using component harnesses are run in. For unit and integration tests using test runners such as Karma, Jasmine, or Jest, we use `TestbedHarnessEnvironment`, which is bundled with the Angular CDK. For Protractor end-to-end tests, we use `ProtractorHarnessEnvironment`, which is also released as part of the Angular CDK.

> **Important Note**
>
> Protractor support is either deprecated or removed, depending on your Angular version.

If you want to use component harnesses with other end-to-end testing frameworks, you will have to extend the `HarnessEnvironment` base class and implement the `TestElement` interface to work in a different testing environment. Of course, first make sure to look for existing solutions in the Angular ecosystem.

Only one harness environment can be active at any time. We use the harness environment to create a harness loader. A harness loader has the context of a certain DOM element and is used to query for and create component harnesses based on selectors.

We will walk through simple code examples featuring harness environments and harness loaders after discussing the API of component harnesses that are distributed as part of Angular Material.

Component harnesses

A component harness can technically represent any DOM element and a set of user interactions and traits.

To get the feel for a component harness, let's first look at the testing harness for the Angular Material Button component.

The following is the API of `MatButtonHarness`, which is not inherited from the common component harness base classes:

- `blur(): Promise<void>;`
- `click(relativeX: number, relativeY: number): Promise<void>;`

  `click('center'): Promise<void>;`

  `click(): Promise<void>;`
- `focus(): Promise<void>;`
- `getText(): Promise<string>;`
- `isDisabled(): Promise<boolean>;`
- `isFocused(): Promise<boolean>;`

It has a few other methods that it inherits from an Angular CDK base class, but we will not discuss those for now.

I trust that you can guess what the methods do, based on their names, parameters, and return values. Notice that they are all asynchronous, that is, they all return a `Promise`.

Every method except `getText` represents a user interaction. The `getText` method reads content from the DOM, which is displayed to the user, the text of the button to be precise.

Next, let's explore the API of the testing harness for the Angular Material Select component.

The following is the API that is specific to `MatSelectHarness`:

- `blur(): Promise<void>;`
- `clickOptions(filter?: OptionFilters): Promise<void>;`

  Pick the drop-down option(s) matching the specified filter. For multi-option selects, multiple options can be picked. For single-option selects, the first matching option is picked.
- `close(): Promise<void>;`

  Closes the drop-down panel.
- `focus(): Promise<void>;`
- `getOptionGroups(filter?: OptionGroupFilters): Promise<OptionGroup[]>;`

  Read drop-down option groups matching the specified filter.
- `getOptions(filter?: OptionFilters): Promise<Option[]>;`

  Read drop-down options matching the specified filter.
- `getValueText(): Promise<string>;`

  Read the value of the chosen drop-down option.
- `isDisabled(): Promise<boolean>;`
- `isEmpty(): Promise<boolean>;`

  Resolves `false` if no value has been picked. Resolves `true` if a value has been picked.
- `isFocused(): Promise<boolean>;`
- `isMultiple(): Promise<boolean>;`

  Resolves `true` if the harness wraps a multi-option select component. Resolves `false` if it wraps a single-option select component.
- `isOpen(): Promise<boolean>;`
- `isRequired(): Promise<boolean>;`

- `isValid(): Promise<boolean>;`
- `open(): Promise<void>;`

These methods correspond to our expectations about the behavior and information represented by a drop-down picker such as the Angular Material **Select** component.

Now that we have discussed the most important concepts of a component harness and seen a few component harness APIs, it is time to look at a test case that combines these concepts.

The following is an example of a test for an online clothing store. You will have to imagine the implementation of the `ShirtComponent` and collaborating services. In fact, this is what the test-as-a-user approach is all about. It is component- and implementation-agnostic:

1. First, we import the necessary Angular packages:

```
import { HarnessLoader } from '@angular/cdk/testing';
import { TestbedHarnessEnvironment } from '@angular/cdk/
testing/testbed';
import { TestBed } from '@angular/core/testing';
import { MatButtonModule } from '@angular/material/
button';
import { MatButtonHarness } from '@angular/material/
button/testing';
import { MatSelectModule } from '@angular/material/
select';
import { MatSelectHarness } from '@angular/material/
select/testing';
```

2. Next, we import the `ShirtComponent` component, the collaborating `OrderService`, and the `OrderSpyService` class to replace it:

```
import { OrderService } form './order.service';
import { OrderSpyService } form './order-spy.service';
import { ShirtComponent } from './shirt.component';
```

3. Before we can implement test cases, we configure the Angular testing module by setting up the necessary declarables and replacing `OrderService` with a spy service for testing purposes:

```
describe('ShirtComponent', () => {
beforeEach(() => {
    TestBed.configureTestingModule({
declarations: [ShirtComponent],
      imports: [MatButtonModule, MatSelectModule],
      providers: [
        { provide: OrderService, useClass:
         OrderSpyService },
      ],
    });
    const fixture = TestBed.createComponent(
     ShirtComponent);
```

4. We use the component fixture from the previous step to create a harness loader for unit tests:

```
    loader = TestbedHarnessEnvironment.
     loader(fixture);
```

5. Finally, we store the order spy service in the shared `orderSpy` variable:

```
    orderSpy = TestBed.inject(OrderService) as
     OrderSpyService;
});
  let loader: HarnessLoader;
  let orderSpy: OrderSpyService;
```

6. Now we load the component harness for the shirt size picker, which is implemented using an Angular Material Select component as seen in this step:

```
it('orders a Large shirt', async () => {
const shirtSizePicker = await loader.
getHarness(MatSelectHarness);
```

7. For this test case, we also have to load a component harness for the **purchase** button, which is implemented using the Angular Material Button component:

```
      constpurchaseButton = awaitloader.getHarness(
       MatButtonHarness.with({ text: '1-click purchase' });
```

8. Next, we perform a sale as the user by picking a `Large` shirt and clicking the **purchase** button:

```
  await shirtSizePicker.clickOptions({ text: 'Large' });
      await purchaseButton.click('center');
```

9. Finally, we assert that the order service spy has been called as expected:

```
      expect(orderSpy.purchase).
       toHaveBeenCalledTimes(1);
    });
  });
```

We see that the test is relatively straightforward because of it using Angular Material's component harnesses. First, the *large* shirt size is picked, then the one-click **purchase** button is clicked, and we expect our order service spy to have been called with an order.

Notice how we start out by configuring the Angular testing module as usual in a component test. After creating the component fixture, we use it to create a harness loader. The harness loader is then used to query for component harnesses for Angular Material Select and Button components. A filter is passed to make sure we interact with the correct button.

We use the component harnesses instead of interacting with component instances or passing selectors to DOM queries. Our test is decoupled from structural DOM changes and implementation details of Angular Material's components.

Now that we have explored the most important concepts of component harnesses, you are ready to implement your own and use them to test as a user in *Part 2, Build a Real-World Application with the Angular Ivy Features You Learned.*

Summary

In this chapter, we have explored the APIs of the Angular YouTube Player, the Google Maps Angular components, the Angular CDK's Clipboard API, as well as the Angular CDK's component harnesses and how they are used by Angular Material, which we in turn can use in our applications.

The YouTube Player component is an Angular-specific wrapper around the embedded YouTube Player. We learned how to initialize it and explored its API in detail.

Many official Angular component wrappers are available to create and interact with the rich API of Google Maps. We learned about the `GoogleMap`, `MapMarker`, `MapMarkerClusterer`, and `MapInfoWindow` components, which are used for common **Geographic Information System** (**GIS**) use cases.

The Angular CDK's Clipboard API is a cross-browser and cross-platform compatible API for interacting with the native clipboard. We learned about the `CdkCopyToClipboard` directive, the `Clipboard` service, and the `PendingCopy` class.

Finally, we discussed the main concepts of Angular component harnesses introduced by the Angular CDK. We saw examples of component harness APIs exposed by Angular Material and how we can use them to test our own components without relying on implementation details or DOM structures, which might change in a future version of the package.

With all these exciting new features and APIs fresh in our minds, let's move on to *Part 2, Build a Real-World Application with the Angular Ivy Features You Learned*, in which we add new functionality to an existing Angular application. Surely, this knowledge will come in handy.

Chapter 5, Using CSS Custom Properties, starts off *Part 2, Build a Real-World Application with the Angular Ivy Features You Learned* by combining CSS Custom Properties and Angular to add a theme picker to the Angular Academy application.

5
Using CSS Custom Properties

We will now dive into the practical details of how to implement the features you encountered in *Part 1 A Quick and Functional Guide to Angular Ivy* in the Angular Academy application, which will allow you to browse and select a curriculum of available Angular video courses on YouTube.

We'll start by describing the Angular Academy application while covering the following topics:

- Building a theme picker using custom CSS properties
- Implementing the theme service
- Controlling CSS Grid templates using custom CSS properties

By the end of the chapter, you will have gained practical experience in using custom CSS properties in a real-world application.

Technical requirements

Combining and using all the new features in the Angular Academy application will involve a series of practical choices that are usually hidden in the descriptions in *Part 1, A Quick and Functional Guide to Angular Ivy* which means we will need to provide an overview of the application as a whole.

So, before we dive into an in-depth description of all the details, let's check out the code first and have it available for inspection while reading the in-depth description in the upcoming chapters.

Open the terminal and issue the following command:

```
git clone https://github.com/PacktPublishing/Accelerating-Angular-Development-with-Ivy
```

The demo project source code will be placed in `projects/demo` and can be started on your development machine, like this:

```
cd Accelerating-Angular-Development-with-Ivy
npm install
ng serve demo
```

If you go to `http://localhost:4200` on your browser, you should now be able to see the Angular Academy app:

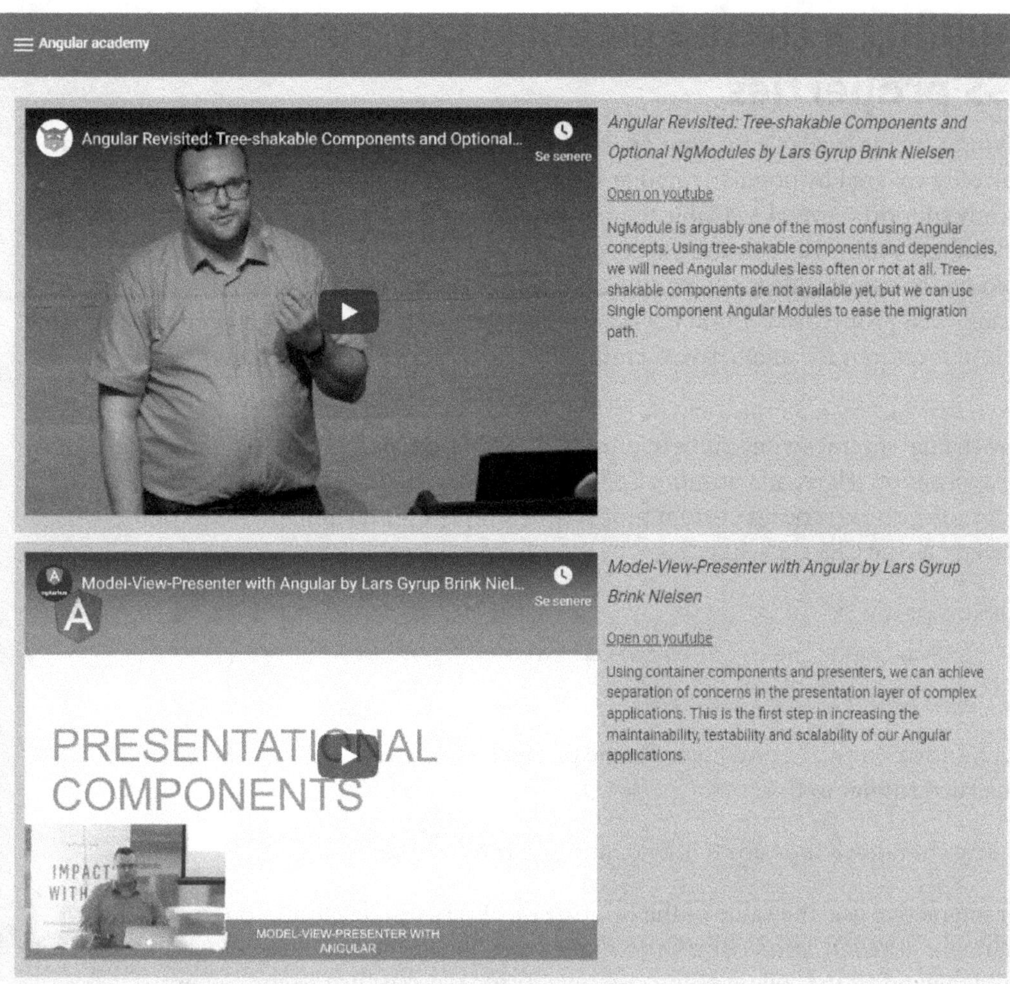

Figure 5.1 – In the Angular Academy application, you can see a list of YouTube video courses

When the Angular Academy application is started, you will see a list of YouTube courses on the default page (as illustrated in *Figure 5.1*). Feel free to browse to get a feel of what we will be covering in the upcoming chapters. As an example, try clicking the **Edit Theme** menu entry to open the theme picker. We will be looking at this in the next section.

Building a theme picker using custom CSS properties

Having a theme available for your application is a common and important use case that is already covered by popular Angular libraries. You may be aware that Angular Material already supports several available themes (for example, the popular **deeppurple-amber** and **indigo-pink**). The common approach of using preprocessor variables when using SCSS has been available for some time. But now, you can support dynamic theming using custom CSS properties without having to generate the CSS files using a preprocessor. This adds new options for interactive theming that we will cover in this section.

Given that further CSS rules can be built using CSS Custom Properties, we can now change several styling rules in one go, directly from one of the components in the application. Here, one or more CSS classes can be calculated on the fly via the `--headerbackground` custom property or simply by attaching the value of the property as the CSS class, like this:

```
.mycomponent {
  background: var(--headerbackground, white);
}
```

The `headerbackground` color can be picked by the user using a theme picker construct, similar to this:

```
<input name="headerBackground" type="color" />
```

For interactive use, the value of the `headerbackground` custom property could be a variable stored in `localStorage` that's available during use. `headergroundcolor` could then affect the styling of the elements inside the specific tile component.

In the Angular Academy application, we will wrap it with an Angular Material form field in the theme component template, like this:

```
<mat-form-field appearance="fill">
    <mat-label> Header background </mat-label>
    <input
     matInput
     name="headerBackground"
     (blur)="update($event)"
     type="color"
     [value]="headerBackground"
```

```
    />
</mat-form-field>
```

Using the Material form input field in the theme picker component will look like this:

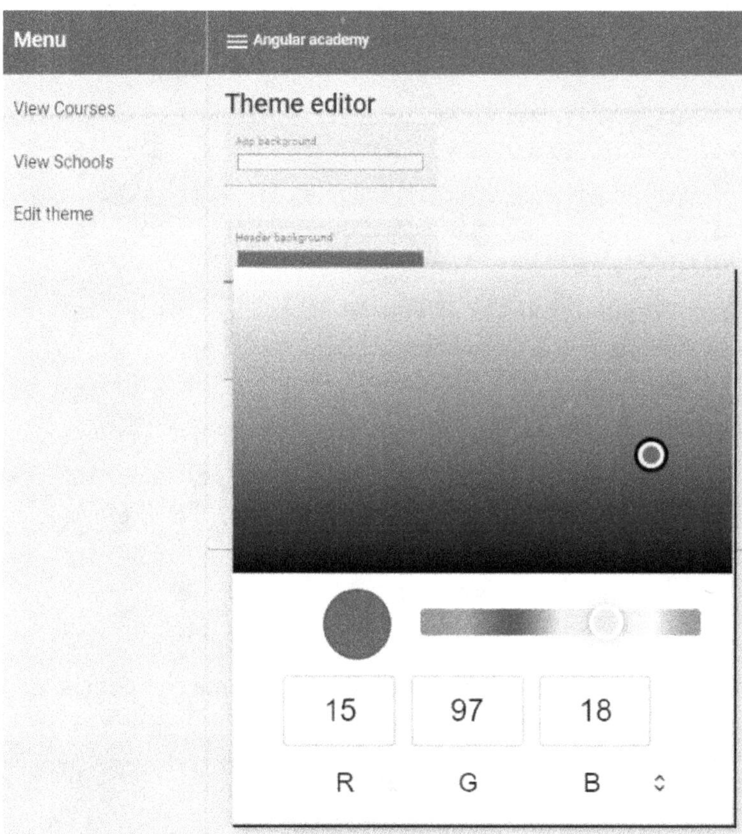

Figure 5.2 – You can use the theme picker component to select the header's background color

We can apply the chosen color *directly* from the custom SCSS properties to the relevant component SCSS files in the application or directly at the app scope using the style syntax precedence works implemented in Ivy.

For practical purposes, we will use a `@HostBinding` to bind each theme setting to a theme variable for the application scope, like this:

```
export class AppComponent {
  @HostBinding('style.--background')
  background: string;

  @HostBinding('style.--headerbackground')
  headerBackground: string;

  @HostBinding('style.--tilebackground')
  tileBackground: string;
  constructor(themeService: ThemeService) {
    this.background = themeService.getSetting(
     'background');
    this.headerBackground = themeService.getSetting(
     'headerBackground');
    this.tileBackground = themeService.getSetting(
     'tileBackground');
  }
}
```

The values for the CSS properties can be retrieved in the theme component via a getter, which retrieves the value via the theme service, like this:

```
  get headerBackground(): string {
    return this.themeService.getSetting(
     'headerBackground');
  }
```

The `update($event)` callback on the theme component could refer to the theme service, which would update the chosen value behind the scenes. Additionally, the theme component would allow access to the chosen values for the custom CSS properties by binding them to the application's scope, like this:

```
  update(event: any): void {
    this.themeService.setSetting(event.target.name,
     event.target.value);
  }
```

In the app component, we are referring to the same theme service for handling data updates related to the theme component. By separating handling data updates into a service, we can abstract the details of how to store and retrieve the theme settings for later.

Implementing the theme service

The theme service has the responsibility of retrieving the dynamic theme settings.

Let's start with a simple implementation by using `localStorage`. In this implementation, we will also provide default settings if the value is not available:

```
import { Injectable } from '@angular/core';
@Injectable({
  providedIn: 'root',
})
export class ThemeService {
  constructor() {}
  public setSetting(name: string, value: string): void {
    localStorage.setItem(name, value);
  }
  public getSetting(name: string): string {
    switch (name) {
      case 'background':
        return localStorage.getItem(name) || 'yellow';
      case 'tileBackground':
        return localStorage.getItem(name) || '#ffcce9';
      case 'headerBackground':
        return localStorage.getItem(name) || '#00aa00';
    }
    return 'white';
  }
}
```

Here, we will provide example settings if the setting has no previous value. For example, if `headerBackground` has not previously been set, then we will set it to `#00aa00`.

One of the benefits of using the theme service from the theme picker component is that you, at a later stage, could choose to implement the theme service to use another mechanism for storing and retrieving settings (for example, you could choose to retrieve values from a corporate design token system that may contain the default settings). Additionally, you can also use a provider scope for the service to share data related to the usage scenario, as you will see later in *Chapter 8, Additional Provider Scopes*. Another aspect of theming is controlling the relative sizes and placement of items on the screen, depending on user preferences. A modern approach to this is using CSS Grids – and it turns out that we can encapsulate these settings well using custom CSS properties. We will cover how to do this in the next section.

Controlling CSS Grid templates using custom CSS properties

Imagine that you want to emphasize the importance of the textual descriptions of videos, so you would like to increase the amount of space for text, and most likely decrease the amount of space used for the video. This could be implemented using a dynamic viewer using some TypeScript logic, which could perform the sizing calculations on the fly. Given that you would like to be able to view the content on your mobile phone as well, you would need to implement considerations for the grid layout on smaller screens. This additional requirement would be complex enough that custom theming would be required. However, as it turns out, we can combine media queries with inline custom CSS properties in a manner that is both compact and easy to understand.

If we introduce the `video` and `text` CSS classes for the course video tiles, then we can style them using custom CSS properties and CSS Grid techniques while referencing the `container` grid-columns, like this:

```
.tile {
  background: var(--tilebackground, grey);
  padding: 15px 15px 15px;
  overflow: hidden;
  &.video {
    grid-column: span var(--videosize, 9);
  }
  &.text {
    grid-column: span var(--textsize, 3);
  }
}
```

The `videosize` and `textsize` CSS properties would then control how many columns are assigned to `video` and `text`. The declaration of the `container` grid looks like this:

```
.container {
  display: grid;
  grid-template-columns: repeat(12, 1fr);
  grid-template-rows: 1fr;
  grid-auto-flow: dense;
  padding: 15px 15px 15px;
  align-content: center;
}
```

These new CSS classes could be combined in the `Course` component to render the videos alongside accompanying text descriptions, like this:

```
<div class="container">
  <div class="tile video">
    <p>
      <youtube-player videoId="{{ videoId }}"></youtube-
        player>
    </p>
  </div>
  <div class="tile text">
    <h3>
      <i>{{ title }}</i>
    </h3>
    <p>
      <a href="https://youtube.com/watch?v={{ videoId }}">
        Open on youtube</a>
    </p>
    <p>{{ description }}</p>
  </div>
</div>
```

To control the sizing inside the course tiles, we will introduce two new variables to the theme component: **video size** and **text size**. These two variables will be bound in their range by the available columns in the CSS Grid (in this case, 12). Additionally. it would make sense that they sum up to the number of columns.

The **video size** and **text size** sliders could be implemented as part of the theme picker. This would look like this:

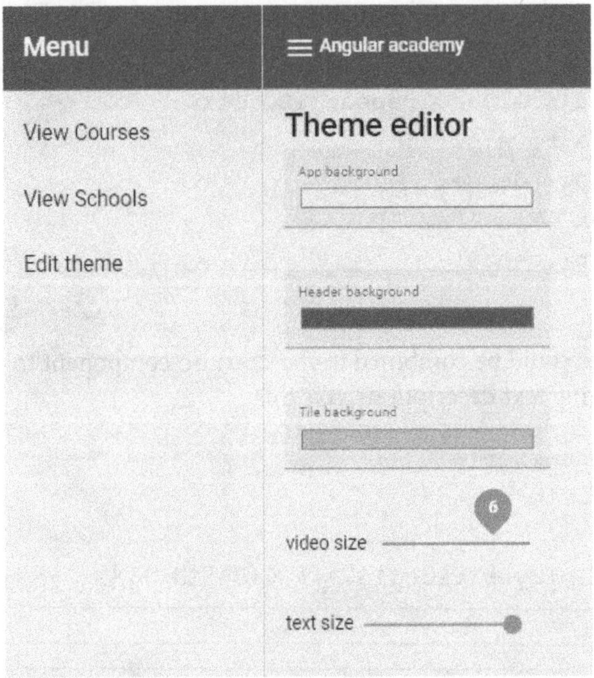

Figure 5.3 – Using sliders for video and text size

For practical purposes, we will bound the **video size** variable between `3` and `7` (to allow for a max of `5` for **text size**). This should leave sufficient room for both the video and the text inside the grid. This simple approach can be implemented via a Material Slider, like this:

```
<mat-label> video size </mat-label>
<mat-slider
  thumbLabel
  min="3" max="7" step="1"
  (input)="setSize('videoSize', $event)"
  [value]="videoSize"
>
</mat-slider>
```

Here, the `setSize` callback will update the variable using the theme service in the same manner as the other variables – just with the added complexity that we have received a `MatSliderChange`:

```
setSize(name: string, event: MatSliderChange): void {
  this.themeService.setSetting(name,
   event.value?.toString() || '1');
  location.reload();
}
```

Every time we change the value in one of the sliders, the layout of components inside the grid could be changed. We should pay attention when including output from external components that have their own layout systems (for example, the YouTube Player). We will learn about integrating the YouTube Player in the next chapter, but for now, let's just settle with a simple `location.reload()`. This should illustrate how to force a render of the grid (and all the other components too).

Before we leave the topic of using custom CSS properties from Angular, let's remember that we want to support smaller screens as well. It turns out that we can introduce an elegant solution by combining custom CSS properties and media queries without referring to Angular code, like this:

```
.tile {
  background: var(--tilebackground, grey);
  padding: 15px 15px 15px;
  overflow: hidden;
  @media screen and (min-width: 768px) {
    &.video {
      grid-column: span var(--videosize, 9);
    }
    &.text {
      grid-column: span var(--textsize, 3);
    }
  }
  @media only screen and (max-width: 768px) {
    grid-column: span 12;
  }
}
```

Here, we can see one of the strengths of adding support for using custom CSS properties directly from within Angular in the application scope: by using a combination of CSS Custom Properties and media queries outside the Angular components, we can rely on the expertise of dedicated designers that can work with little or no knowledge about Angular code. Additionally, we would also have the option of creating integrations that could retrieve theme settings from corporate design token systems, without requiring the application to be redeployed when the design tokens change.

Summary

In this chapter, you encountered the Angular Academy application for the first time and learned how CSS Custom Properties can be used to implement the theme picker using a theme service. Along the way, you also had a brief first encounter with the YouTube Player in the course list.

In the next chapter, we will dig deeper into how to use the YouTube Player and how you can combine a curriculum of YouTube videos using the Angular components you learned about in *Part 1, A Quick and Functional Guide to Angular Ivy*.

6
Using Angular Components

In the previous chapter, we introduced the Angular Academy application and briefly scratched its surface by demonstrating how we can use CSS Custom Properties to control theme properties and CSS Grid layouts. In this chapter, we will dig into the details of how to implement the application using the new Angular components that we introduced in *Chapter 4, Exploring Angular Components Features*.

Note that we will be covering a lot of ground in this chapter, so I recommend that you revisit *Chapter 4, Exploring Angular Components Features*, if you want to learn about Angular components or if you need a refresher. This chapter will be about using the new official Angular components and how to wire them together in our Angular Academy example application.

More specifically, we will be covering the following topics in this chapter:

- Understanding the Angular Academy application
- Showing course videos using the Angular YouTube Player
- Using the new Clipboard service
- Finding your school using the Angular Google Maps components

If you are like us, then you are probably eager to dive into the details on how to use the new official Angular components, but let's step back and reflect on the Angular Academy application so that we can understand what we are going to build.

By starting from the ground up in this way, we hope that you get a more concrete understanding of how to use the new components by seeing how to wire them together using services and navigation.

Understanding the Angular Academy application

The first thing you should think about when using components is **what** you want the user to be able to accomplish by using the component, as well as what data that use case requires, before digging into the details about the component you are going to use.

Our primary use case for the Angular Academy application. is to allow tailor-made lists of video courses, which will provide the user with a custom list of videos and location information for the video's content. To accomplish this, we will use the Angular YouTube Player to display videos and select schools via Angular Google Maps (for the **how**). It will be the school that creates the course, and the course will contain one or more videos that the user will watch.

With the use case in place, we can now give some thought to **what** data we will be using, before describing **how** we are going to display or use the data. This will make it easier to separate data retrieval and storage from the actual usage of the different components. So, let's establish a data model before we use the new Angular components.

Establishing a data model

We have many ways to describe data models, but for the sake of brevity, we will limit ourselves to simple TypeScript interfaces to describe the models and use Angular services to communicate with the backend.

The models that we will be using to support our use case are as follows:

- School
- Course
- Video

The list of videos you saw in the previous chapter is related to a Course that has been produced by a School. We'll start by giving a Course a title, an optional description, and a list of videos for the user to watch:

```
export interface ICourse {
  id: string;
  title: string;
  description?: string;
  videos: IVideo[];
}
```

The basic information we need for each video is where to access it on YouTube, the date it was uploaded, the author who produced it, and an optional description of what it is about:

```
export interface IVideo {
  externalId: string;  // YouTube ID
  title: string;
  date: Date;
  author?: string;
  description?: string;
}
```

We will attach a name and the latitude and longitude for each school so that we can find them on a map:

```
export interface ISchool {
  id: string;
  name: string;
  longitude: number;
  latitude: number;
  courses: ICourse[];
}
```

Additionally, we will describe the courses being offered by the school in the `courses` array. Note that the description in the `courses` array will be using the Course model as part of the School model. The shared use of the Course interface will allow logic to be reused between the Schools and Course components.

Dividing the application into components

Now, we will divide the app into the following three main modules. These will be displayed separately on the screen:

- Course
- Theme
- Schools

The Course component will use the layout grid that we introduced in *Chapter 5, Using CSS Custom Properties,* as well as a Video component that will display YouTube videos via the Angular YouTube Player module.

The theme component should also seem familiar to you from *Chapter 5, Using CSS Custom Properties,* where we used it to control the theme settings using CSS properties. The theme settings should influence the graphical display of the Schools and Course components.

The Schools component will let you find your school via the Angular Google Maps component and let you select a course to follow from the chosen school (and redirect you to the Course component).

Each of the components will be mapped for navigation in `app.module.ts`, like this:

```
const routes: Routes = [
  { path: '', redirectTo: 'course/1', pathMatch: 'full' },
  { path: 'course/:id', component: CourseComponent },
  { path: 'schools', component: SchoolsComponent },
  { path: 'theme', component: ThemeComponent },
];
```

Note that the Course component will require a parameter for which course to display. For simplicity, we will just assume that the user is already logged in, has chosen to follow the course with an ID of `1`, and then display this as the default route. A later implementation could add a login page and a user model, which could redirect the user to the chosen course upon startup (the chosen course could, for example, be stored on the user model). We will revisit this in *Chapter 8, Additional Provider Scopes.*

Now that we have divided the application into components, it is time to start thinking about how to include dependencies for these modules.

Including dependencies using modules

We will start by specifying the application as a module that imports the Course, Schools, and Theme modules:

```
@NgModule({
  bootstrap: [AppComponent],
  declarations: [AppComponent, NavigationComponent],
  imports: [
    CommonModule,
    BrowserAnimationsModule,
    RouterModule.forRoot(routes, { initialNavigation:
      'enabledNonBlocking' }),
    LayoutModule,
    CourseModule,
    SchoolsModule,
    ThemeModule,
    MaterialModule,
  ],
})
export class AppModule {}
```

The Course module will include the Video module:

```
@NgModule({
  declarations: [CourseComponent],
  imports: [CommonModule, ThemeModule, VideoModule,
   MaterialModule],
  exports: [VideoModule],
})
export class CourseModule {}
```

The Video module will include the `YouTubePlayerModule` and `ClipboardModule` dependencies:

```
@NgModule({
  declarations: [VideoComponent],
  imports: [
    CommonModule
    YouTubePlayerModule,
    ClipboardModule,
  ],
  exports: [VideoComponent],
})
export class VideoModule {}
```

Finally, the Schools module will include `GoogleMapsModule`:

```
@NgModule({
  declarations: [SchoolsComponent],
  imports: [CommonModule, MaterialModule,
   GoogleMapsModule],
})
export class SchoolsModule {}
```

Did you notice that we only included the specific dependencies where they are needed? This way of structuring applications can help you have a clearer overview of the dependencies in the application.

Retrieving data using services

Now that we have specified the example data model and divided the application into modules, it is time to specify how we will access data from the components. We will use the following Angular services for this:

- `CourseService`
- `SchoolsService`

Each of the services will be set up to fetch data asynchronously. The main difference here will be that `CourseService` will retrieve one Course at a time, while `SchoolsService` will retrieve several Schools at a time.

The `CourseService` will have a `getCourse` call available to retrieve a single course:

```
@Injectable({
  providedIn: 'root'
})
export class CourseService {
  constructor() { }
  getCourse(courseId: string): Observable<ICourse> {
    return of(mockCourse);
  }
}
```

The Course model will contain the list of videos for the course to display.

> **Important Note**
> We are using mock data here, but this approach should illustrate how you can implement asynchronous data retrieval from the server via the `Course` component.

Similarly, we will introduce `SchoolsService`, which will retrieve the list of schools providing courses:

```
@Injectable({
  providedIn: 'root'
})
export class SchoolsService {
  constructor() { }
  getSchools():  Observable<ISchool[]> {
      return of(mockSchools);
  }
}
```

Here, we will return several schools via the `getSchools` call. The idea is that each of the returned schools should provide links to one or several courses that can be retrieved via `CourseService`. We will illustrate this by providing a link to the courses that the school is offering in the Schools component.

Wiring up navigation

For simplicity, we will assume that the user has signed up for a school so that we can direct the user to a course in the chosen school. Here, the default link will display videos from the course in the chosen school via the default route on startup.

We will start by setting up the links for the Material Side Navigation container in the Navigation component on the left side of the screen, as follows:

```
    <mat-nav-list>
        <a mat-list-item href="/#">Watch course</a>
        <a mat-list-item href="/schools">Find school</a>
        <a mat-list-item href="/theme">Edit theme</a>
   </mat-nav-list>

```

Here, you can see that the default route, /, will have a title of **Watch courses** This will be mapped to the Courses component in the route description.

Having established the navigation and the data model and having split the app into modules, we can start describing how to use the Angular components in the Angular Academy app. We will start by describing how to use the Angular YouTube Player to show the course videos.

Showing course videos using the Angular YouTube Player

In this section, we'll create a separate video component to display videos that are attached to a course. For simplicity, we'll accept the information as `@Input` inside the Video component, like so:

```
@Component({
  selector: 'workspace-video',
  templateUrl: './video.component.html',
  styleUrls: ['./video.scss'],
})
export class VideoComponent implements OnDestroy, OnInit {
  private youtubeIframeScript: HTMLScriptElement;

  @Input()
```

```
  public title!: string;

  @Input()
  public name!: string;
  @Input()
  public videoId!: string;
  @Input()
  public description!: string;
  @Input()
  public snippet!: string;

  get youtubeLink () {
   return this.title
     + ": https://www.youtube.com/watch?v="+this.videoId;
  }
  constructor(@Inject(DOCUMENT) private document: Document) {
    this.youtubeIframeScript =
     this.document.createElement('script');
    this.youtubeIframeScript.src =
     'https://www.youtube.com/iframe_api';
    this.youtubeIframeScript.async = true;
  }
  ngOnInit(): void {
    this.document.body.appendChild(
     this.youtubeIframeScript);
  }

  ngOnDestroy(): void {
    this.document.body.removeChild(
     this.youtubeIframeScript);
  }
}
```

This code should be familiar to you from the introduction in *Chapter 4, Exploring Angular Components Features,* where we introduced how to use it. We can now write the template for the Video component to display the YouTube videos, like this:

```
<div class="container">
  <div class="tile video">
    <p>
      <youtube-player videoId="{{ videoId }}"></youtube-
       player>
    </p>
    <p>
      <button [cdkCopyToClipboard]="youtubeLink">Copy video
       link to clipboard</button>
    </p>
  </div>
  <div class="tile text">
    <h3>
      <i>{{ title }}</i>
    </h3>
    <p>{{ description }}</p>
 </div>
</div>
```

Do you remember how we used the custom `videoSize` CSS property in the previous chapter to tweak the grid column's size? This effort is paying off – we just need to refer to the `video` class here (with no direct reference to dynamic sizing at all). The reference to the `tile` class lets us manipulate the tile color settings using the theme component.

Did you also notice how we introduced the `cdkCopyToClipboard` feature here? This feature can be handy in desktop applications when you want to extract data from the running application to the clipboard.

Having established the Video component, we can now use it from the Course component, like this:

```
<ng-container *ngIf="course$ | async as course">
  {{ course.title }}
  <div *ngFor="let video of course.videos">
    <workspace-video
      videoId="{{video.externalId}}"
      title="{{video.title}}"
      description="{{video.description}}"
    >
    </workspace-video>
  </div>
</ng-container>
```

Note the usage of the `async` pipe operator on `course$`. Here, we are waiting for the data to be retrieved so that we can start rendering the video using the Video component.

Now that we have covered how to display course videos using the Angular YouTube Player, we will learn how to find a school using the Angular Google Maps component and show how the navigation from the School component to the Course component works.

Finding your school using the Angular Google Maps component

The Schools component will allow you to find a school via Google Maps by clicking a marker where the school has been placed. This will open the **MapInfo** window, where you can click on the course that you can watch from school. Clicking on this course will lead you to the course overview you saw in the previous chapter.

You can find the Schools component in the Angular Academy application by clicking the **Find school** menu entry. This should render the Schools component, where you will see the example **Angular Advanced** school.

When you open the Schools component in the Angular Academy application, it should open with a red default Google Maps marker. If you click it, then your display should look like this:

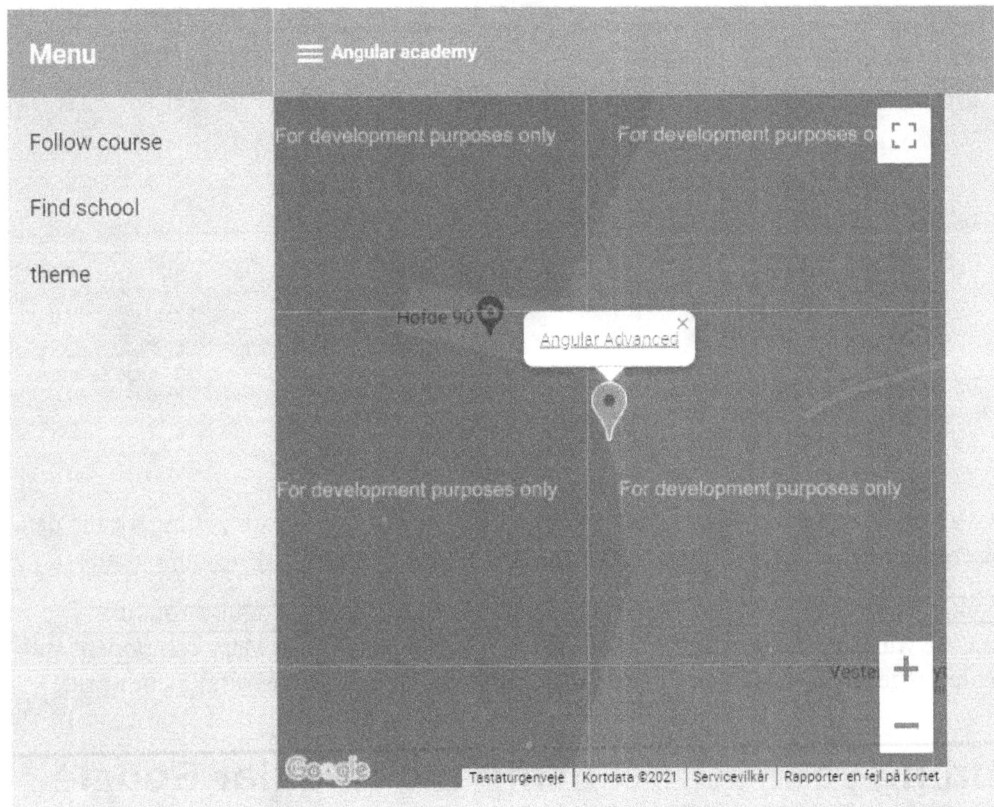

Figure 6.1 – Opening the red Google Maps marker

If you click the **Angular Advanced** link in the map information window, you will be transferred to the course component with the ID for the **Angular Advanced** course as a parameter.

We expect the incoming course data on the map info window of the Schools component to come from the Schools service as an asynchronous call, as illustrated here:

```
@Component({
  selector: 'workspace-schools',
  templateUrl: './schools.component.html',
  styleUrls: ['./schools.component.scss'],
```

```
})
export class SchoolsComponent {
  @ViewChild(GoogleMap, { static: false }) map!: GoogleMap;
  @ViewChild(MapInfoWindow, { static: false }) info!:
   MapInfoWindow;
  school!: ISchool;
  schools$: Observable<ISchool[]>;

  constructor(
    schoolsService: SchoolsService
  ) {
    this.schools$ = schoolsService.getSchools();
  }
  openInfo(anchor: MapAnchorPoint, school: ISchool): void {
    this.school = school;
    this.info.open(anchor);
  }
}
```

Here, you can see how we opened `MapInfoWindow` when clicking on the map anchor point in the `openInfo` call.

We will register the `openInfo` call on the `mapClick` function of the Angular Google Maps component, and make it open a **MapInfo** window that displays a link to the courses being offered by the school:

```
<ng-container *ngIf="schools$ | async as schools">
  <google-map [center]="{ lat: 56.783778, lng: 8.228937 }">
    <map-marker
      *ngFor="let school of schools"
      #marker="mapMarker"
      [position]="{ lat: school.longitude, lng:
       school.latitude }"
      (mapClick)="openInfo(marker, school)"
    ></map-marker>
    <map-info-window>
      <div *ngFor="let course of school?.courses">
```

```
        <a href="course/{{ course.id }}"> {{ course.title
         }} </a>
       </div>
     </map-info-window>
   </google-map>
 </ng-container>
```

Note the usage of the async pipe on the `schools$` observable. This will make the school data available in the `schools` variable. Combined with `NgIf` on `<ng-container>`, you can stop data from being rendered until data is available.

If you have not tried it yet, I recommend that you try clicking your way through **Find your school** and clicking the **Angular Advanced** course to find your way back to the Course listing.

Did you notice how the Course listing was opened using the course ID that was retrieved from the course array in the School data model? In our example application, we have registered the `course/:id` route to open the course component with the `id` parameter. For now, we will just assume that the user only has a single course available and that this course is chosen upon startup. This simple example has helped illustrate how the basic flow of the Angular Academy is intended to be. In a more realistic scenario, we would allow for the user to log in and have chosen courses stored in a session. We will revisit this in *Chapter 8, Additional Provider Scopes*, where we will introduce user login.

Summary

In this chapter, we gave you a basic overview of how to use the new Angular components by giving you a concrete example with the Angular Academy application. The example included a description of how to retrieve data using services and how to structure the inclusion of dependencies via Angular modules. In the next chapter, we will describe how to use a test-as-a-user approach using Angular Component Harnesses.

7
Component Harnesses

Testing is a fundamental part of software development. It helps ensure that the code that's delivered covers the feature requirements and is free of implementation issues. Sometimes, when testing UI code, it is hard to avoid performing tightly coupled tests on the DOM structure. However, Angular Ivy brings a new solution for this. Component testing harnesses make it possible to develop a testing API for our components using the industry-standard Page Object pattern, but on a more granular level.

To top this off, component harnesses for the Angular Material directives and components are already included in Angular Ivy. In this chapter, we will learn how to use the component testing harnesses in Angular Components and how to implement custom component harnesses to make it easier to test our components.

We will cover the following topics in this chapter:

- Using Angular Material's component harnesses
- Creating a component harness

By the end of this chapter, you should have gained an overview of how and where to use component harnesses.

Using Angular Material's component harnesses

You saw an example of how to use the **Material Button harness** in *Chapter 4, Exploring Angular Components Features.* Now, let's explore how to test the **theme** component using the material test harnesses with a *Test as a user* strategy.

As you might remember, the theme component lets us select the color and size settings for Angular Academy. The user can do this by selecting a color by going to the **Color Input** field to obtain the wanted color for the given **Header background** setting. To simulate this action while testing as a user, we will use a `MatInputHarness` with the `#headerBackground` selector:

```
it('should be able to read default header background
  color theme setting', async () => {
  const headerBackground: MatInputHarness = await
    loader.getHarness(
      MatInputHarness.with({ selector: '#headerBackground'
      })
    );
    expect(await headerBackground.getValue()
    ).toBe('#00aa00');
});
```

Here, we expect the default setting to be `'#00aa00'`. We retrieve the value from the test harness using the `getValue` method.

In this example, we could also simply find the value in the input field with the `'#headerBackground'` ID and check its value. So, let's build a more complicated test where we should be able to change the header background color theme setting:

```
it('should be able to change the header background color
  theme setting', async () => {
    const headerBackground: MatInputHarness = await
      loader.getHarness(
       MatInputHarness.with({ selector: '#headerBackground'
        })
    );
```

```
    headerBackground.setValue('#ffbbcc').then(() => {
      expect(themeService.getSetting(
        'headerBackground')).toBe('#ffbbcc');
    });
  });
```

As we did earlier, we will use the component harness to interact with the **Material Input** button for the header background input button. Now, we will set the value to `'#ffbbccc'` and check that this setting has been picked up by the theme setting.

Did you notice that we did not write `fixture.detectChanges()` in this test? We can avoid this because we are using the component harness that will handle the DOM operations here. Clicking this input field and interacting with the color selector as the user would have done is somewhat complex using DOM operations, but here, we are using the operations on the component testing harness instead. By doing this, we can avoid brittle changes in the test related to change detection.

Similarly, we can use `MatSliderHarness` to `MatInputHarness` to avoid performing DOM operations when testing the **Video Size** slider setting for the theme component:

```

  it('should be able to check default text and video slider
   settings', async () => {
    const videoSizeSetting =
     Number(themeService.getSetting('videoSize'));
    expect(videoSizeSetting).toBe(7);
    const videoSizeSlider: MatSliderHarness = await
     loader.getHarness(
      MatSliderHarness.with({ selector: '#videoSizeSlider'
       })
    );

    expect(await videoSizeSlider.getId()
     ).toBe('videoSizeSlider');
    expect(await videoSizeSlider.getValue()
     ).toBe(Number(videoSizeSetting));
```

Now, we can retrieve the default `videoSize` setting from the theme service and the screen component to check that it is `7` by using API operations from the test harness. Using the `async/await` construct in combination with the test harness leads to fairly compact code here.

By now, you should know how to use the existing material component harnesses. Next, let's dive into how to build component harnesses for the Angular Academy app.

Creating a component harness

Let's imagine that we want to expose our `Video` component so that it can be integrated with other applications. Here, it would make sense to write a test harness for it – but how should we structure it? Our example of displaying YouTube videos using a YouTube Player component inside a `Video` component, which will be inside a Course component, turns out to be difficult to test using a "test as a user" approach in the DOM directly. So, let's take a layered approach here.

When constructing a component, we should strive to only have a single reference point – the DOM – for each page. Taking a layered approach, we start the test from the Course component, which knows about the Video component, which, in turn, knows about the YouTube Player component. By doing this, we can test from the `Course` component by exposing the `Video` harness that encapsulates the `workspace-video` selector as a Page Object for each of the instances of the `Video` component, like this:

```
export class VideoHarness extends ComponentHarness {
  static hostSelector = 'workspace-video';

  protected getTextElement = this.locatorFor('.text');
  async getText(): Promise<string|null> {
    const textElement = await this.getTextElement();
    return textElement.text()
  }
  textEquals(video: IVideo, text: string): boolean {
    return
      text?.toLowerCase().trim().includes(
         video.title.trim().toLowerCase()
      );
  }
}
```

Now, we can display the text for each video by using the `getText()` call from the `course.component.spec.ts` file. Then, we can use the supplied `textEquals` call to test for equality:

```
  it('should render the video title in text when displaying it',
 async () => {
    const renderedVideos =
      await loader.getAllHarnesses(VideoHarness);
    courseService.getCourse('1').subscribe((course) => {
        renderedVideos.forEach(async (video: VideoHarness) => {
          const text = await video.getText()|| "";
          expect(course.videos.find((v) =>
           video.textEquals(v, text))).toBeTruthy();
        });
      });
  });
```

Here, we looped through all the rendered videos for course `'1'` in the course component and checked that each of the texts accompanying the rendered videos contain the course title we can retrieve via the course service. Note that the `textEquals` function is supplied from the test harness here, which means we can change that function in a later version of the component library.

We will let the `Video` harness hide the DOM operations related to the videos using a Page Object approach. The `Video` harness will then know about the implementation that's specific to the `YouTubePlayer` harness, which encapsulates the `youtube-player` selector, like this:

```
class YoutubePlayerHarness extends ComponentHarness {
  static hostSelector = 'youtube-player';
  async getVideoId(): Promise<string|null> {
    const host = await this.host();
    return await host.getAttribute('ng-reflect-video-id');
  }
}
```

When rendering videos using the Angular YouTube Player, we expect the video IDs to be available. The trick here is that we want to hide the implementation details for the Angular YouTube Player from the test for the Course component. So, let's introduce the `getVideoId` function to the `Video` harness so that it is available when we test from the Course component:

```
it('should have the videoId available when rendering the
 video', async() => {
  const renderedVideos = await
   loader.getAllHarnesses(VideoHarness);
  renderedVideos.forEach( async(video: VideoHarness) => {
    const videoId = await video.getVideoId();
    expect(videoId).toBeTruthy();
  })
})
```

The full `Video` harness will look like this:

```
export class VideoHarness extends ComponentHarness {
  static hostSelector = 'workspace-video';

  protected getTextElement = this.locatorFor('.text');
  protected getVideoElement =
   this.locatorFor(YoutubePlayerHarness);

  async getText(): Promise<string | null> {
    const textElement = await this.getTextElement();
    return textElement.text();
  }

  async getVideoId(): Promise<string | null> {
    const videoElement = await this.getVideoElement();
    return videoElement.getVideoId();
  }

  textEquals(video: IVideo, text: string): boolean {
     return  text?.toLowerCase().trim().includes(
      video.title.trim().toLowerCase());
```

```
    }
}
```

At this point, we can expose the `Video` test harness, along with the `Video` component, to anyone that wants to use our `Video` component in their application.

Summary

In this chapter, we looked at some examples of how to use some of the existing test Material UI test harnesses in the context of the Angular Academy application. Additionally, we introduced how to implement a component harness for the `Video` component that is used in the Angular Academy application.

In the next chapter, we will wrap up our Angular Academy application by showing you how to use the new provider scopes.

8
Additional Provider Scopes

This chapter seeks to explain how to use dependency injection scopes to develop more lean components and features in Angular Ivy. To explore these features, we will learn how to create a non-singleton service and how to reuse dependencies across Angular elements.

We will introduce the `any` provider scope by revising the theme service so that it can accept specific configurations when used in different scenarios using the `any` provider scope and rewiring the schools and course modules to be lazy loaded.

We will then wrap up *Part 2, Build a Real-World Application with the Angular Ivy Features You Learned* by building a new login element that shows how to share information across application boundaries by using the platform provider scope for Angular Elements.

We will cover the following topics in this chapter:

- Revisiting the root provider scope
- Using the any provider scope for a configurable theme service
- Sharing information across application boundaries using the platform provider scope

Before we dive into the details about the new provider scopes, let's take a moment to reflect on the services we have introduced so far using the root provider scope.

Revisiting the root provider scope

So far, we have discussed the following services in the Angular Academy application using the root scope provider:

- `SchoolsService`: Retrieve information about the available schools.
- `CourseService`: Retrieve information about the course.
- `ThemeService`: Set and retrieve information about the current theme.

These services have served us well as singletons in the app so far – and marking the services for use with `providedIn: 'root'` via the `Injectable` decorator makes it pretty easy to use them for the standard use case. If you have been around since the early days of Angular, then you might have been accustomed to injecting services as dependencies in each specific module – for example, you might have been wondering why `SchoolsService` is not listed in the providers array from the schools module here:

```
@NgModule({
  declarations: [SchoolsComponent],
  imports: [CommonModule, GoogleMapsModule],
  exports: [SchoolsComponent],
})
export class SchoolsModule {}
```

We do not need to insert explicit providers here as we have had tree-shakable providers since Angular version 6. We can now just rely on the `injectable` decorator. This makes the Angular modules a bit leaner and easier to configure, and we can provide alternative implementations for the service later.

Providing singleton services on the root scope sounds useful in itself (it worked well in *Chapter 6, Using Angular Components*). But what if we want to have specific service instances per use case? It turns out that we can do that by using the any provider scope for the theme service and changing the modules to be lazy loaded instead of the default eager loading. Let's dig into the details on how to do that.

Using the any provider scope for a configurable ThemeService

Let's use the any provider scope for a configurable `ThemeService` by injecting configurable settings depending on the use case for each module that we load:

```
@Injectable({
  providedIn: 'any',
})
export class ThemeService {
  constructor(@Inject(themeToken) private theme: ITheme) {}

  public setSetting(name: string, value: string): void {
    this.setItem(name, value);
  }

  public getSetting(name: string): string {
    switch (name) {
      case 'background':
        return this.getItem(name) ?? this.theme.background;
      case 'tileBackground':
        return this.getItem(name) ??
         this.theme.tileBackground;
      case 'headerBackground':
        return this.getItem(name) ??
         this.theme.headerBackground;
      case 'textSize':
        return this.getItem(name) ?? this.theme.textSize;
      case 'videoSize':
        return this.getItem(name) ?? this.theme.videoSize;
    }
    return 'white';
  }

  private setItem(name: string, value: string): void {
    localStorage.setItem(this.prefix(name), value);
```

```
  }

  private getItem(name: string): string | null {
    return localStorage.getItem(this.prefix(name));
  }

  private prefix(name: string): string {
    return this.theme.id + '_' + name;
  }
}
```

We introduced the theme service in *Chapter 5, Using CSS Custom Properties*. Let's make that configurable by introducing an `InjectionToken` instance for the theme:

```
import { InjectionToken } from '@angular/core';
import { ITheme } from './theme/theme.model';

export const themeToken = new InjectionToken<ITheme>('theme');
```

The theme token holds configuration settings that implement the `ITheme` interface:

```
export interface ITheme {
  id: string;
  background: string;
  headerBackground: string;
  tileBackground: string;
  textSize: string;
  videoSize: string;
}
```

We could then use a `green` theme with these values in `AppModule` via the `InjectionToken` token of `theme`:

```
import { ITheme } from './app/theme/theme.model';

export const theme: ITheme = {
  id: 'green',
  background: '#f8f6f8',
```

```
    tileBackground: '#f4ecf4',
    headerBackground: '#00aa00',
    textSize: '3',
    videoSize: '7',
  };
```

Note that the configured settings will only be the starting values. The user can still change them while the system is running.

Now we can inject the theme settings into the injector scope using the theme service. When we are using the any provider scope, we can obtain an instance for every lazy-loaded module that injects the service. Here is a snippet that shows the relevant parts of how to rewire the modules for lazy loading via an app routing module and while running the `green` theme in the `AppModule`:

```
// (...)
import { themeToken } from './theme.token';
import { theme } from '../green.theme';

import { AppRoutingModule } from './app-routing.module';
// (...)
@NgModule({
  declarations: [AppComponent],
  imports: [
    // (...)
    AppRoutingModule,
  ],

  providers: [
    {
      provide: themeToken,
      useValue: theme,
    },
  ],
  bootstrap: [AppComponent],
})
export class AppModule {}
```

Here we provide default settings for `ThemeService` on the application module scope. The dependencies for each module would be dynamically loaded from inside `AppRoutingModule` like this:

```
const routes: Routes = [
  {path: '', redirectTo: 'login', pathMatch: 'full',},
  {
    path: 'course',
    loadChildren: () =>
      import('./course/course.module').then((m) =>
       m.CourseModule),
  },
  {
    path: 'login',
    loadChildren: () =>
      import('./login/login.module').then((m) =>
       m.LoginModule),
  },
  {
    path: 'schools',
    loadChildren: () =>
      import('./schools/schools.module').then((m) =>
       m.SchoolsModule),
  },
  {
    path: 'theme',
    loadChildren: () =>
      import('./theme/theme.module').then((m) =>
       m.ThemeModule),
  },
];
```

Then we need to establish a routing module for each module. As an example, the routing module for the course module looks like this:

```
import { CourseComponent } from './course.component';

const routes: Routes = [{ path: ':id', component:
CourseComponent }];

@NgModule({
  imports: [RouterModule.forChild(routes)],
  exports: [RouterModule],
})
export class CourseRoutingModule {}
```

We already have the settings defined in the `Provider` scope for the app module that uses the course module, so there is no need to redefine it here if we want the `green` theme for a module. But if we want to use another theme, then we can introduce another theme configuration via the `theme` token like this:

```
import { ITheme } from './app/theme/theme.model';

export const theme: ITheme = {
  id: 'metallic',
  background: '#ffeeff',
  tileBackground: '#ffefff',
  headerBackground: '#ccbbcc',
  textSize: '3',
  videoSize: '7',
};
```

We could then use the `metallic` theme on the lazy-loaded `LoginModule` via the platform injector scope like this:

```

// (...)
import { theme } from '../../metallic.theme'
// (...)

@NgModule({
```

```
  declarations: [LoginComponent],
  imports: [
    CommonModule,
    MaterialModule,
    FormsModule,
    ReactiveFormsModule,
    LoginRoutingModule,
  ],
  providers: [
    {
      provide: themeToken,
      useValue: theme,
    },
  ],
})
export class LoginModule {}
```

Since we are lazy loading the login module, we will now create a new instance of the theme service – so that the `login` component can use the `metallic` theme instead of the `green` theme like the rest of the application. In this way, we can use an instance of the theme service to render the toolbar using the `green` theme and the `login` component using the `metallic` theme, like this:

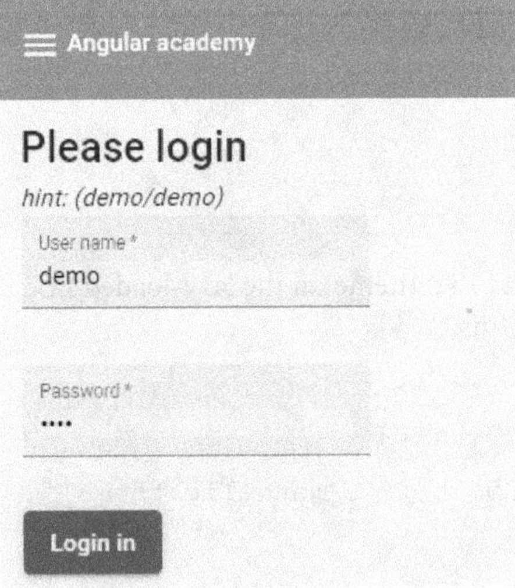

Figure 8.1 – The login screen. Note that the background here is from the metallic theme

This will be the first screen you see when you start the Angular Academy application. Note that the metallic card background is set in the `login.component.scss` file using the mechanism you learned about in *Chapter 5, Using CSS Custom Properties*:

```
.mat-card {
  background: var(--background, green);
}
```

The `background` variable will be set in `LoginComponent` like this:

```
@Component({
  selector: 'workspace-login',
  templateUrl: './login.component.html',
  styleUrls: ['./login.component.scss'],
})
export class LoginComponent {
  @HostBinding('style.--background')
  background: string;

  loginForm: FormGroup;

  constructor(
    public fb: FormBuilder,
    public authService: AuthService,
    private themeService: ThemeService
  ) {
    this.background = themeService.getSetting(
     'background');
    this.loginForm = this.fb.group({
      name: [''],
      password: [''],
    });
  }

  loginUser(): void {
    this.authService.login(this.loginForm.value);
  }
}
```

`LoginComponent` passes user information from `loginForm` to `AuthService`:

```
@Injectable({
  providedIn: 'platform',
})
export class AuthService {
  public loginEvent: EventEmitter<string> = new
   EventEmitter();

  login(user: IUser): void {
    if (user.name === 'demo' && user.password === 'demo') {
      this.token = 'thisTokenShouldBeProvidedByTheBackend';
      this.loginEvent.emit(this.token);
    }
  }

  logout(): void {
    this.token = '';
  }

  set token(value: string) {
    localStorage.setItem('token', value);
  }

  get token(): string {
    return localStorage.getItem('token') ?? '';
  }
}
```

Note that we are using a very simple way of obtaining a `login` token for the demo user with the password demo here. This example could be extended to call the backend of your choice and submit `LoginEvent` after doing so.

The idea is that we can react to `LoginEvent` like this:

```
this.authService.loginEvent.subscribe((token: string) => {
    const courseId: string | null =
     this.preferenceService.getCourseId();
    if (courseId) {
      router.navigate(['/course', courseId]);
    } else {
      router.navigate(['/schools']);
    }
  });
```

By now we hope that you have had the chance to play around with the Angular Academy application to see how things are wired together. Did you notice that you are redirected directly to the course in the Angular Academy application after logging in a second time? The first time you log in, you should be given the opportunity to choose a school from a map – and then choose a course from the school. When you choose a course, this will be stored via the preference service. This preference can then be used to redirect users to courses.

The Angular Academy application uses some fairly complex navigation logic that is only relevant inside the app – but what if we wanted to share information outside of the app? We can do exactly that by using the `platform` provider scope in combination with Angular elements.

Sharing information across application boundaries using the platform provider scope

We can demonstrate how to share information outside of the app by creating a tweet button as an Angular element. This Angular element could be used outside of the app as well. Let's dig into the details on how to do that.

First, we will start by adding Angular elements to the app by running the following command:

```
ng add i @angular/elements
```

Then we include the Twitter widgets SDK within the page like this:

```
<script
    src="https://platform.twitter.com/widgets.js"
    type="text/javascript"
  ></script>
```

Then we can build a tweet hashtag button using a `TweetCourse` component like this:

```
<ng-container *ngIf="course$ | async as course">
  <a
    href="https://twitter.com/intent/tweet?button_hashtag=
     {{ course.hashtag }}"
    class="twitter-hashtag-button"
    >Tweet {{ course.hashtag }}</a
  >
</ng-container>
```

The `TweetCourse` component uses `CourseService` with the platform provider scope to retrieve data:

```
@Component({
  selector: 'workspace-tweetcourse',
  templateUrl: './tweetcourse.component.html',
})
export class TweetCourseComponent implements OnInit {
  @Input()
  courseId!: string;

  public course$: Observable<ICourse> | undefined;
  constructor(
    private courseService: CourseService
  ) {
  }
  ngOnInit(): void{
    this.course$ = this.courseService.getCourse(
     this.courseId);
```

```
  }
}
```

Now, we register `TweetCourseComponent` as an Angular element like this:

```
export class AppModule implements DoBootstrap {
  constructor(private injector: Injector) {
    const el = createCustomElement(TweetCourseComponent, {
      injector: this.injector,
    });
    customElements.define('tweet-course', el);
  }
  ngDoBootstrap(): void {}
}
```

It can then be used as a web component:

```
  <tweet-course courseId="1"></tweet-course>
```

Since `CourseService` is registered on the platform provider scope, we can now use it both from within our new `<tweet-course>` Angular element and inside our Angular Academy application.

We insert the `<tweet-course>` element inside the navigation bar for the desktop version of the application like this inside the navigation component:

```
    <mat-nav-list>
      <a *ngIf="!token" mat-list-item
       href="/login">Login</a>
      <ng-container *ngIf="token">
        <a mat-list-item *ngIf="courseId"
         routerLink="course/{{ courseId }}"
          >Follow course</a
        >
        <a mat-list-item routerLink="/schools">Find
         school</a>
        <a mat-list-item routerLink="/theme">Theme</a>
        <a *ngIf="token" mat-list-item (click)="logout()"
         href="#">Logout</a>
      </ng-container>
```

```
      <ng-container *ngIf="courseId">
        <tweet-course courseId="courseId"></tweet-course>
      </ng-container>
    </mat-nav-list>
```

Here you see that the `<tweet-course>` element should be rendered in the sidebar if you have chosen a course (if you have a `courseId` instance). It should look like this:

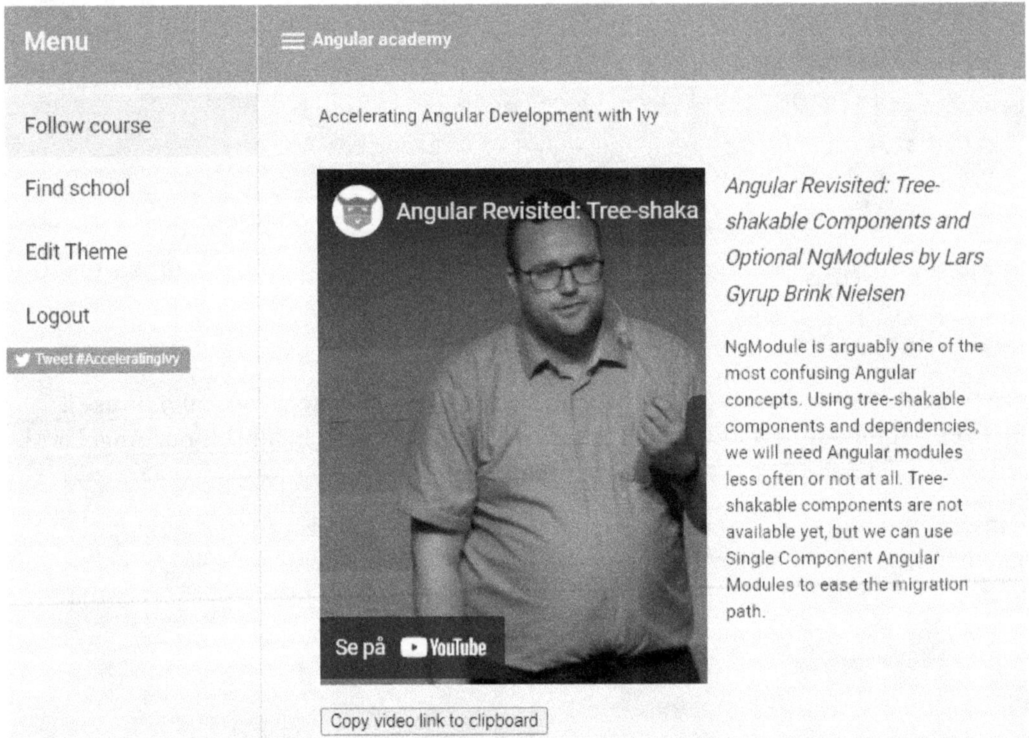

Figure 8.2 – Introducing the Tweet #AcceleratingIvy button

If you are logged in to Twitter when you press the **Tweet #AcceleratingIvy** button, then your Twitter should open and let you tweet using the `#AcceleratingIvy` hashtag. If you register another hashtag for your course in the course service, then this hashtag will be presented.

Maybe you have other ideas for components that could be used outside of the application? Did you notice that we marked `AuthService` as `providedIn: 'platform'`? You could export the `Login` component as an Angular element and update it to integrate it into your own application platform to perform single sign-on.

Summary

In this chapter, we started by expanding on the usage of the root provider scope and introduced the new any and platform provider scopes in the context of the Angular Academy application. We then introduced the any provider scope by applying lazy loading via `AppRoutingModule`, which allowed us to use a separate theme for `LoginModule`. Finally, we saw how to create a Tweet button that can be used with the platform provider scope.

In the next chapter, we will start on Part 3, Upgrade Your View Engine Application and Development Workflow to Angular Ivy of the book and look into migrations and more practical aspects of using Angular Ivy.

9
Debugging with the New Ivy Runtime APIs

Angular Ivy introduces a new API for inspecting and debugging our Angular applications at runtime. It replaces the previous `NgProbe` API and allows tree-shaking of `DebugElement`.

We will explore Angular's most useful runtime debugging functions, including the following:

- `ng.applyChanges`
- `ng.getComponent`
- `ng.getContext`
- `ng.getListeners`

Having these debugging utilities at hand will allow you to verify your assumptions about active components, their templates, and their DOM bindings at runtime.

This chapter covers these topics:

- Introduction to the new Ivy runtime API
- Inspecting an active component
- Inspecting event listeners
- Inspecting an embedded view context

Getting familiar with these topics will improve your development workflow when implementing Angular Ivy applications.

Technical requirements

To support all the features used in the code examples of this chapter, your application requires at least the following:

- Angular Ivy version 12.0
- TypeScript version 4.2

Also, note that the runtime debugging API is only available when Angular runs in development mode.

You can find complete code examples for the random number generator in this book's companion GitHub repository at `https://github.com/PacktPublishing/Accelerating-Angular-Development-with-Ivy/tree/main/projects/chapter9/random-number`.

Introducing the new Ivy runtime API

If you have worked with Angular versions before Angular Ivy, you might be familiar with the `NgProbe` API, which was available in the global scope at runtime as the `ng.probe` function. Angular Ivy replaces this API with a set of new runtime debugging functions, which are only available in Angular development mode.

The new API contains the following functions:

- `ng.applyChanges(component: {}): void;`

  Mark the specified component for dirty checking if it is using the `OnPush` change detection strategy. Afterward, trigger a change detection cycle.
- `ng.getComponent<T>(element: Element): T | null;`

  Resolve the Angular component that is attached to the specified DOM element.

- `ng.getContext<T>(element: Element): T | null;`

  When passed a DOM element generated by a structural directive such as `NgIf` or `NgFor`, resolve the view context of the embedded view. In other cases, resolve the parent component.

- `ng.getDirectiveMetadata(directiveOrComponentInstance: any): ComponentDebugMetadata | DirectiveDebugMetadata | null;`

  Resolve the metadata of the specified Angular component or directive instance.

- `ng.getDirectives(element: Element): {}[];`

  Resolve Angular directives – but not components – attached to the specified DOM element.

- `ng.getHostElement(componentOrDirective: {}): Element;`

  Resolve the host element that the specified component or directive is attached to.

- `ng.getInjector(elementOrDir: {} | Element): Injector;`

  Resolve the injector associated with the specified element, directive, or component.

- `ng.getListeners(element: Element): Listener[];`

  Resolve event listeners attached to the specified DOM element. This does not include host listeners created from directive or component metadata but includes event listeners not added by Angular.

- `ng.getOwningComponent<T>(elementOrDir: {} | Element): T | null;`

  Resolve the host component of the specified DOM element, directive, or component.

- `ng.getRootComponents(elementOrDir: {} | Element): {}[];`

  Resolve the root components associated with the specified DOM element, directive, or component, that is, the components bootstrapped by Angular.

The `Listener` data structures returned by `ng.getListeners` have the following interface:

```
interface Listener {
  callback: (value: any) => any;
  element: Element;
  name: string;
  type: 'dom' | 'output';
```

```
  useCapture: boolean;
}
```

The data structures returned by `ng.getDirectiveMetadata` have the following interfaces:

```
interface DirectiveDebugMetadata {
  inputs: Record<string, string>;
  outputs: Record<string, string>;
}
interface ComponentDebugMetadata extends DirectiveDebugMetadata
{
  changeDetection: ChangeDetectionStrategy;
  encapsulation: ViewEncapsulation;
}
```

The `inputs` and `outputs` properties defined in the preceding debug metadata interfaces contain object maps from data binding property names to component property names.

It is important to be aware that the `ComponentDebugMetadata#changeDetection` and `ComponentDebugMetadata#encapsulation` properties are number enumerations, so their values will be numbers at runtime, not strings. This makes them a bit harder to interpret when debugging.

As seen in the preceding overview, most of these runtime debugging utilities accept as a parameter a DOM element, a component instance, or a directive instance. From the specified object, they look up DOM elements, one or more component instances, one or more directive instances, an injector, or event listeners attached to the DOM.

The one that stands out is `ng.applyChanges`. We will discuss when and how to use it in the next section.

Inspecting a component instance

To explore our application programmatically at runtime, we often need a reference to an active component instance. Once we have a component reference, we can change bound properties and call event handlers or other methods.

However, first, we need a reference to either a directive instance or a DOM element with an attached component. Using the **Elements** tab of our browser developer tools, we can pick a DOM element and the developer tools will store a reference in the global `$0` variable. Alternatively, we can use `document.querySelector`, or any other DOM querying or traversing API.

Say we have a component that generates random numbers, as seen in the following figure:

Random number generator

Your random number is 590

GENERATE

Figure 9.1 – A component for generating a random number

It has a component model as seen in the following code block:

```
import { Component } from '@angular/core';

@Component({
  selector: 'app-random-number',
  templateUrl: './random-number.component.html',
  styleUrls: ['./random-number.component.css'],
})
export class RandomNumberComponent {
  generatedNumber?: number;

  onNumberGenerated(generatedNumber: number): void {
    this.generatedNumber = generatedNumber;
  }
}
```

In its template, it is using Angular Material's Button component as seen in the following code listing:

```
<ng-container
  #generator="appRandomNumber"
  appRandomNumber
  (numberGenerated)="onNumberGenerated($event)"
```

```
  ></ng-container>

<mat-card>
  <mat-card-header>
    <mat-card-title>Random number generator</mat-card-
     title>
  </mat-card-header>

  <mat-card-content>
    <p>Your random number is {{ generatedNumber }}</p>
  </mat-card-content>

  <mat-card-actions>
    <button mat-button (click)= "generator.
generateNumber()">GENERATE</button>
  </mat-card-actions>
</mat-card>
```

Given that we have a reference to the `<button>` DOM element in the `$0` variable, we can do the following to resolve two different component instances:

```
ng.getComponent($0);
// -> MatButton
ng.getOwningComponent($0);
// -> RandomNumberComponent
```

There is a subtle but important distinction between `ng.getComponent` and `ng.getOwningComponent`. The first invocation returns an instance of the `MatButton` component, which is attached to the `<button>` DOM element. The second invocation gives us a reference to the active instance of `RandomNumberComponent`.

We conclude that `ng.getComponent` returns the component attached to the specified DOM element, in this case `MatButton`. Now, `ng.getOwningComponent` returns the component instance that's associated with the component template used to generate the specified DOM element, in this case an instance of `RandomNumberComponent`.

The `generatedNumber` UI property is bound to text in the DOM created for the random generator component by Angular. What if we wanted to change it to something specific, say `42`? With a reference to an active component instance, we can change UI properties directly as seen in the following browser console listing:

```
const component = ng.getOwningComponent($0);
component.generatedNumber = 42;
```

However, when looking at the rendered application, we notice that the DOM has not been updated to reflect this new component state. When using the runtime debugging API, we must let Angular know when we have manually changed the state and want Angular to update the DOM it manages.

We notify Angular of the dirty state by passing the component instance to `ng.applyChanges` as follows:

```
ng.applyChanges(component);
```

After Angular completes a change detection cycle, we notice that the new state is reflected in the DOM, as seen in the following figure:

Random number generator

Your random number is 42

GENERATE

Figure 9.2 – A component displaying a manually specified number

Great, now you are familiar with the most common runtime debugging functions, which put us in control of Angular components. In the following sections, we will look at debugging APIs that are instrumental to runtime debugging, yet not as commonly used.

Inspecting event listeners

We'll reuse the random number generator example from the *Inspecting a component instance* section.

For reference, the following is the random number directive used in the template of the random number component:

```
import {
  Directive, EventEmitter, OnInit, Output
```

```
} from '@angular/core';
@Directive({
  exportAs: 'appRandomNumber',
  selector: '[appRandomNumber]'
})
export class RandomNumberDirective implements OnInit {
  #generatedNumber?: number;
  @Output()
  numberGenerated = new EventEmitter<number>();

  ngOnInit(): void {
    this.generateNumber();
  }

  generateNumber(): void {
    this.#generatedNumber =
      Math.floor(1000 * Math.random());
    this.numberGenerated.emit(this.#generatedNumber);
  }
}
```

When its `generateNumber` method is called, it outputs the randomly generated number through the `numberGenerated` output property. The random number component has bound its event handler `RandomNumberComponent#onNumberGenerated` to this custom component event.

DOM event listeners

We pick the `<button>` element in the **Elements** tab of our browser developer tools, so that a reference to it is stored in `$0`.

Notice in the component template that the button component has a click event binding. We want to access it so that we can trigger it. To do that, pass the button DOM element to the `ng.getListeners` function, as seen in the following browser console listing:

```
const [onButtonClick] = ng.getListeners($0);
// onButtonClick -> Listener
onButtonClick.callback();
```

We unpack the first and only event listener that is returned by `ng.getListeners` when passed the button DOM element. This `Listener` data structure is stored in the `onButtonClick` variable.

We invoke the click event handler by invoking `onButtonClick.callback`. This triggers the same state update as clicking the **Generate** button. However, Angular is not aware of the dirty state.

> **Important Note**
>
> DOM event listeners registered outside of Angular code are also returned by `ng.getListeners`.

You might remember from the *Inspecting a component instance* section that we must notify Angular of state changes we introduce through the runtime debugging APIs. We do this by passing the component instance to `ng.applyChanges` as follows:

```
const component = ng.getOwningComponent($0);
ng.applyChanges(component);
```

When Angular finishes a change detection cycle, the newly generated number is displayed in the DOM for the random number generator component, which is managed by Angular.

Notice that we did not pass any parameters to the `Listener#callback` method. In our use case, the event handler did not accept any parameters. If it did, we would most likely have to pass arguments of the expected type for it to work. For example, a click event listener might accept a `MouseEvent` event of type `click`.

Custom component event bindings

Event handlers bound by Angular custom component event bindings are also registered as listeners. Our example component has an inline event handler bound for the custom `numberGenerated` event.

We pick the `<span>` element in the **Elements** tab of our browser developer tools so that a reference to it is stored in `$0`.

We pass the span element to `ng.getListeners`, and notice that it lists two listeners, one of type `"dom"` and another one of type `"output"`, as shown here:

```
const [domListener, outputListener] = ng.getListeners($0);
// domListener -> Listener { element: span, name:
"numberGenerated", useCapture: false, type: "dom", callback: f
}
```

```
// outputListener -> Listener { element: span, name:
"numberGenerated", useCapture: false, type: "output", callback:
f }
```

We simulate a random number generation by passing `7` to `outputListener.callback` and run change detection by passing the component to `ng.applyChanges`. This is shown in the following browser console listing:

```
const component = ng.getOwningComponent($0);
outputListener.callback(7);
ng.applyChanges(component);
```

Once a change detection cycle has finished, the random number generation that we simulated is displayed in the DOM, which is managed by Angular.

That is all for inspecting native and custom event listeners using the Angular runtime debugging API. In the final section of this chapter, we'll learn what an embedded view context is, and how to inspect it using Angular Ivy's runtime debugging API.

Inspecting an embedded view context

A structural directive is used to add and remove elements to the DOM throughout the lifecycle of a component. They create an embedded view, which is bound to a view context. This is the case of the `NgIf` and `NgFor` directives that are part of the Angular framework.

> **Important Note**
>
> Only one structural directive can be attached to an element. If you need to apply multiple structural directives, wrap the element in the special `<ng-container>` element, attach the outer structural directive to this element, and so on.

When we pass an element with a structural directive attached to `ng.getContext`, it returns the view context. For example, when we pass an element with an `NgIf` directive attached to it, `NgIfContext` is returned, which has the following shape:

```
interface NgIfContext {
  $implicit: boolean;
  ngIf: boolean;
}
```

The embedded view that is dynamically created by `NgIf` is bound to the `$implicit` property of `NgIfContext`.

If we instead pass an element that has an `NgFor` directive attached to it, `NgForOfContext` is returned. It has the following shape:

```
interface NgForOfContext<T> {
  $implicit: T;
  count: number;
  index: number;
  ngForOf: T[];
  even: boolean;
  first: boolean;
  last: boolean;
  odd: boolean;
}
```

An embedded view is dynamically created by the `NgFor` directive for each item in `NgForOfContext#ngForOf`. Each embedded view is bound to the `$implicit` property of `NgForOfContext` specific to that item.

However, each embedded view also has access to the other properties that are specified in `NgForOfContext`. For example, we can loop through a list of users and store a reference to the `index` and `first` context properties, like so:

```
<ul>
  <li *ngFor="let user of users; index as i; first as
   isFirst">
    {{i}}/{{users.length}}.
    {{user}} <span *ngIf="isFirst">(default)</span>
  </li>
</ul>
```

The embedded view created for each user has access to and uses the `index` and `first` properties, which are aliased as `i` and `isFirst`, respectively.

Passing a list item element from the previous code listing to `ng.getContext` results in an `NgForOfContext` value, such as the examples in the following code listing:

```
const listItems = document.querySelectorAll('li');
ng.getContext(listItems[0]);
// -> NgForOfContext { $implicit: "Nacho", count: 4, index: 0, ngForOf: ["Nacho", "Santosh", "Serkan", "Lars"], even: true, first: true, last: false, odd: false }
ng.getContext(listItems[1]);
// -> NgForOfContext { $implicit: "Santosh", count: 4, index: 1, ngForOf: ["Nacho", "Santosh", "Serkan", "Lars"], even: false, first: false, last: false, odd: true }
```

Similarly, if we pass the `<span>` element to `ng.getContext`, we get an `NgIfContext` value, such as the following:

```
ng.getContext(document.querySelector('span'));
// -> NgIfContext { $implicit: true, ngIf: true }
```

Take special care to pass the element that has a structural directive attached, or you will instead receive the closest component instance.

Now you know how to inspect the embedded view context of template elements with a structural directive attached.

Summary

I hope you like the new shiny tools that we added to your toolbox in this chapter. We started with an overview of Angular Ivy's runtime debugging API, which is available only in development mode.

Next, we learned how to inspect a component instance using `ng.getComponent` and `ng.getOwningComponent`. We also changed the component state, then updated the DOM using `ng.applyChanges`.

In the *Inspecting event listeners* section, we used `ng.getListeners` to inspect both native DOM event listeners and custom component event listeners. We passed arguments to their callbacks and triggered change detection using `ng.applyChanges`.

Finally, you now know what an embedded view context is and how to inspect it, for example, how one is created and bound to each component or element managed by the `NgFor` directive. Similarly, we explored an example of an embedded view context for an element managed by an `NgIf` directive.

With all these newfound skills, you are ready to debug Angular applications by inspecting and updating state directly or through events, then reflecting the changes by triggering change detection.

You are even able to inspect the ever-so-hard-to-locate embedded view contexts. Amazing!

In the next chapter, you will learn about the Angular Compatibility Compiler, and when and why it is needed. We will explore its configuration options and optimize it for a CI/CD workflow.

10
Using the Angular Compatibility Compiler

Angular Ivy replaces the previous-generation Angular compiler and rendering runtime known as Angular View Engine. The last version to support the View Engine runtime is Angular version 11.2.

In this chapter, we are going to learn about the bridge between View Engine-compiled Angular packages on npm and your Angular Ivy application, namely the **Angular Compatibility Compiler** (**ngcc**).

This chapter covers the following topics:

- Introducing the Angular Compatibility Compiler
- Using the Angular Compatibility Compiler
- Improving the Angular Compatibility Compiler in your CI/CD workflow

If your Angular Ivy application consumes View Engine-compiled libraries from a package registry, you must use the Angular Compatibility Compiler. After learning about the topics covered in this chapter, you will know what is happening in your local development workflow and be able to fine-tune the Angular Compatibility Compiler in your CI/CD workflow.

Technical requirements

For the techniques demonstrated in this chapter, your application requires at least the following:

- Angular Ivy version 11.1
- TypeScript version 4.0

Introducing the Angular Compatibility Compiler

The source code of Angular libraries is compiled before it is published on a package registry such as npm. Until Angular version 12.0, it was not possible to compile Angular libraries using partial Angular Ivy compilation; they had to be compiled with the View Engine compiler. As part of a transition period, Angular uses the Angular Compatibility Compiler to allow Angular Ivy applications to use libraries that are compiled using the View Engine compiler and published to a package registry.

As of Angular version 12.2, the Angular Compatibility Compiler is still included as part of the Angular CLI, meaning that our Angular Ivy applications can consume libraries that are compiled using either the View Engine or the Angular Ivy compiler.

In Angular CLI version 12.0, partial Ivy compilation for Angular libraries was introduced. In short, it compiles all Angular-specific code except component templates. However, partial Ivy compilation breaks backward compatibility for libraries in that consumers must also have at least Angular CLI version 12.0. In the period following this release, we will see a transition from View Engine-compiled Angular libraries to partially Ivy-compiled Angular libraries.

The good news is that we do not have to change anything in our Angular Ivy application except keep Angular packages up to date. As soon as we have at least Angular 12.0, our application supports partially Ivy-compiled Angular libraries through an internal part of the Angular framework known as the Angular Linker.

The Angular Linker is a replacement for the Angular Compatibility Compiler in that it converts a partially Ivy-compiled Angular library bundle to a fully Ivy-compiled library bundle before including it in the compilation of our application.

As such, the Angular Compatibility Compiler will be removed in a version of Angular unknown at the time of writing but later than version 12.2. When this happens, our Angular application will only be able to use partially Ivy-compiled Angular libraries.

Now that you have an overview of what the Angular Compatibility Compiler and the Angular Linker are and awareness of why they are needed, in the next section, we will discuss how to use the Angular Compatibility Compiler.

Using the Angular Compatibility Compiler

In some Angular version 9 releases, we had to run the Angular Compatibility Compiler manually before building, testing, or serving our Angular Ivy application. In later releases, this changed so that the Angular CLI triggers the Angular Compatibility Compiler as needed.

It is still possible to run the Angular Compatibility Compiler manually. In fact, this allows for fine-tuning it to optimal compilation speed.

The Angular Compatibility Compiler needs to run at least once before any of the following:

- Starting a development server
- Executing automated tests
- Building our application

Every time we install a new version of an Angular library or an additional Angular library from a package registry, we must run the Angular Compatibility Compiler again.

Consider running the Angular Compatibility Compiler as part of your Git repository's `postinstall` hook. When using this technique, we do not have to wait the next time we perform one of the actions mentioned in the previous list. While the Angular Compatibility Compiler is running, we are free to change our source code.

Alternatively, use the `--target` option, as described in the *Angular Compatibility Compiler options* section coming up next.

Angular Compatibility Compiler options

The Angular Compatibility Compiler bundles an executable named `ngcc`. When running this command, we can pass the following options:

- `--create-ivy-entry-points`

  Create an `__ivy_ngcc__` subdirectory inside each Angular library package directory. Inside this directory, another subdirectory will be created with the name of the output bundle format, for example, `fesm2015`. Inside the bundle format folder, the Ivy-compiled bundles and source maps will be placed. If this option is not passed, the original bundles will instead be overwritten.

- `--first-only`

  When this option is used in combination with `--properties`, the Angular Compatibility Compiler will only compile the first module format it recognizes in a library package based on the order of the package property names specified by `--properties`.

- `--properties <package-property-names>`

  This option specifies the acceptable library package formats to compile using the Angular Compatibility Compiler. The package property names refer to the properties of the JSON configuration in a library package's `package.json` module declaration.

  Example: `--properties es2015 browser module main`

- `--target <package-name>`

  This option only compiles the specified package.

  Example: `--target @angular/material/button`

- `--tsconfig <tsconfig-path>`

  You can use this option with `--use-program-dependencies` to target a specific project in your Angular workspace.

  Example: `--tsconfig projects/music-app/tsconfig.app.json`

- `--use-program-dependencies`

  You can use this option to decide which library packages you want to compile with the Angular Compatibility Compiler based on the source code in your Angular workspace or project.

A few more options exist but are for exceptional use cases.

Running the Angular Compatibility Compiler manually, for example, after modifying or adding a package dependency, allows us to optimize the compilation speed. When we are triggering compilation of the entire workspace manually, we should generally use the following command:

```
ngcc --first-only --properties es2015 module fesm2015 esm2015
browser main --create-ivy-entry-points
```

The `--first-only` option ensures that only one package format is compiled into an Angular Ivy-compatible package bundle using the `esm2015` package format. The `--properties` option lists the preferred package format. Research has shown that the `es2015` format is generally the fastest package format to compile from a View Engine-compatible bundle to an Angular Ivy-compatible bundle, closely followed by the `module` format. Finally, the `--create-ivy-entry-points` option is generally faster than in-place bundle replacement.

> **Important Note**
>
> If you are using Angular version 9.0 or 11.1, consider leaving out the `--create-ivy-entry-points` option to use in-place bundle replacement. Research has found this option to be slightly faster in these specific versions.

Consider also adding the `--use-program-dependencies` option to only compile packages that are imported by an application. When using this option, we must run the Angular Compatibility Compiler every time we use a package for the first time in our application.

The `--use-program-dependencies` option is especially useful when using Angular CDK and Angular Material because they have many sub-packages that are all compiled individually. Additionally, every Angular CDK and Angular Material sub-package is compiled by default, not just the ones used by an application. This impacts compilation speed significantly.

> **Important Note**
>
> The `ngcc` commands listed in this chapter are meant to be used in pre-defined command listings inside the `scripts` property of `package.json`. To run them from a terminal, prefix them with `npx`, for example, `npx ngcc --create-ivy-entry-points`.

That is all about options and common techniques for your local development workflow. In the following section, you will learn how to optimize the Angular Compatibility Compiler for speed in a CI/CD workflow.

Improving the Angular Compatibility Compiler in your CI/CD workflow

As you can tell from the description of some of the options supported by the Angular Compatibility Compiler, it maintains files inside your application's `node_modules` folder. Depending on your CI environment, caching and restoring the entire `node_modules` folder might be too slow. In this case, cache your package manager's package cache folder instead.

Maybe caching is not enabled at all in your CI/CD workflow. In both cases, we must run the Angular Compatibility Compiler in every CI/CD workflow run. It starts from scratch with the files it manages.

For this use case, we use guidelines described in *Angular Compatibility Compiler options section*. We use the following `postinstall` hook to run `ngcc` in what is overall the fastest combination of parameter options:

```
ngcc --first-only --properties es2015 module fesm2015 esm2015
browser main --create-ivy-entry-points
```

This only compiles a single package format to the Angular Ivy package format and prefers package formats in the order of the formats that are overall the fastest to compile. The package files are compiled to new files in subfolders managed by the Angular Compatibility Compiler rather than replacing the existing, View Engine-compiled package files.

> **Important Note**
>
> Consider leaving out the `--create-ivy-entry-points` option if you are using Angular versions 9.0 or 11.1. Research has indicated that in-place Ivy compilation is faster in these versions.

Running the Angular Compatibility Compiler in a separate step rather than on demand has the benefit of allowing fine-tuning as we just did. Additionally, it allows us to track the time spent on testing or building our application while excluding the View Engine to Angular Ivy compilation time.

Targeting a single application in a monorepo workspace

As discussed in the *Angular Compatibility Compiler options* section, the Angular CDK and Angular Material are examples of Angular library packages with many sub-packages. If we have a monorepo workspace with several Angular applications, perhaps only some of them are using the Angular CDK or Angular Material. Additionally, any single one of these applications is most likely not using every sub-package of the Angular CDK or Angular Material.

We can take this into account if we have a CI or CD job targeting a single application, for example, a test or build job for a particular application. Imagine that we have a monorepo workspace with two Angular applications, one using the Bootstrap UI component library and the other using Angular Material. In a test or build job for the application using Bootstrap, we use the following command in a step after installing package dependencies:

```
npx ngcc --first-only --properties es2015 module fesm2015
esm2015 browser main --create-ivy-entry-points --tsconfig
projects/bootstrap-app/tsconfig.app.json --use-program-
dependencies
```

We are targeting the Bootstrap application by passing the path of its TypeScript configuration file to the `--tsconfig` option and finally we are adding the `--use-program-dependencies` option.

This will save significant compute time in our CI/CD jobs as our CI server will not have to compile any sub-packages of Angular Material.

Even in the case of the application using Angular Material, we can use a similar command to save time because it will only compile the Angular Material sub-packages that are imported by our application instead of all of them. This is shown in the following example command:

```
npx ngcc --first-only --properties es2015 module fesm2015
esm2015 browser main --create-ivy-entry-points --tsconfig
projects/material-app/tsconfig.app.json --use-program-
dependencies
```

In the preceding command, we changed the path passed to the `--tsconfig` option.

Now you have learned the most common optimization techniques for Angular application CI/CD workflows.

Summary

In this chapter, we first discussed how the Angular Compatibility Compiler is a tool needed in the transition phase while Angular library packages are still compiled using the Angular View Engine compiler. The Angular Compatibility Compiler compiles these package bundles into the Angular Ivy format so that they can be used by our Angular Ivy applications.

Additionally, we discussed how recent versions of Angular support partially Ivy-compiled Angular library packages using the Angular Linker, which eventually fully replaces the Angular Compatibility Compiler.

After reviewing the use cases that rely on the Angular Compatibility Compiler, we briefly discussed the most useful options for the `ngcc` command-line tool. Following that, we walked through common optimization techniques using these options.

Finally, this chapter ended by considering how the Angular Compatibility Compiler can be optimized for speed in CI/CD workflows. We discussed solutions for several specific and common use cases.

Now you know how to take advantage of the Angular Compatibility Compiler and you know when and how to optimize it.

In the next chapter, you will be guided on migrating your existing Angular application from View Engine to Ivy. You will learn about automated and manual migrations as well as other considerations when migrating from View Engine to Angular Ivy.

11
Migrating Your Angular Application from View Engine to Ivy

Several Angular feature releases are published every year. Updating our Angular application requires knowledge of the Angular update process, especially when migrating from Angular View Engine to Angular Ivy as there are many differences, most of which are managed by automated Angular migrations.

In this chapter, you will learn about the steps required to update an Angular application, following the *Angular Update Guide's* instructions, how to manage Angular's third-party dependencies, the most useful parameters for the `ng update` command, how the most important automated Angular Ivy migrations change our applications, and how both automated and manual recommended but optional Angular Ivy migrations are applied.

In this chapter, we will cover the following topics:

- Learning the Angular update process
- Performing automated Angular Ivy migrations
- Performing manual Angular Ivy migrations

Technical requirements

The migrations discussed in this chapter apply to applications at or higher than the following:

- Angular Ivy version 12.1
- TypeScript version 4.2

Make sure you have a recent version of Angular CLI installed globally so that you can run the `ng update` command from a terminal.

Learning the Angular update process

Angular CLI gives us a structured approach to update Angular-specific parts of our application. One type of Angular schematics is **migration**, which modifies our application code to comply with breaking changes. Major and minor version releases of Angular often come with migration schematics.

It is recommended to follow the update process, one major version release at a time. For example, if our application is currently using Angular View Engine version 8.2, we update it to Angular Ivy version 9.1 and verify that all aspects are behaving as expected before we take the next step to update from Angular version 9.1 to version 10.2, and so on until we reach the Angular release version we have planned to update to.

The fewer update steps we perform at a time, the easier it is to identify what went wrong when something did not go as planned.

In this section, we will first learn about the Angular Update Guide, an official web app listing step-by-step instructions. After that, we will discuss Angular's third-party dependencies and how their releases affect our Angular application.

The Angular Update Guide

An important tool for the Angular update process is the Angular Update Guide. Located at `https://update.angular.io`, this web app presents step-by-step instructions for updating our Angular application.

To use the Angular Update Guide, we first choose the following:

- Which Angular version we are currently using
- Which Angular version we want to update to
- The complexity of our application
- Whether our application is a hybrid AngularJS and Angular application using `ngUpgrade`
- Whether we are using Angular Material

Even if we are in a hurry, we should select **Advanced** as our **App complexity** and go through all available instructions to make sure we do not miss any recommended migration steps.

After choosing the option that matches our application, we are presented with a checklist of instructions, divided into the following sections:

- **Before updating**
- **During the update**
- **After the update**

It is not always clear what makes a difference regarding whether an instruction is listed in the **During the update** or **After the update** sections. For a pleasant update process, we make sure to follow the instructions in the **Before updating** section before we follow the instructions in the **During the update** section.

The instructions in the **During the update** section must be followed in the order they are listed in because update and migration commands often depend on each other.

The instructions in the **After the update** section contain both required migrations and recommended migrations. Both manual and automated migrations are listed but not necessarily all migrations related to a release; that is, some automated migrations are applied when using the `ng update` command but are not listed in the Angular Update Guide. Similarly, some recommended automated and manual migrations are listed in the Angular documentation but not in the Angular Update Guide.

Managing Angular dependencies

Outside of official Angular packages, Angular only has a few dependencies. The following package dependencies are listed in Angular's `package.json` files:

- **RxJS**
- **tslib**
- **Zone.js**

Historically, the versions of these package dependencies are managed by the Angular update process. However, migrations for breaking changes are not always available. For example, RxJS has no migrations planned for updating from version 6.x to version 7.x.

Zone.js

At the time of writing, Zone.js is still in a prerelease version. Every minor prerelease version contains breaking changes. Typically, migrations are not necessary for our Angular application because we do not use Zone.js directly. Instead, the `NgZone` API wraps Zone.js.

However, we import Zone.js in several of our application files, and Zone.js version 0.11.1 changes its import paths. Angular version 11 offers an automated migration to update Zone.js.

TypeScript

TypeScript does not follow semantic versioning. Every minor release version contains breaking changes. No automated migrations are available for TypeScript so if our application outputs compilation errors after updating Angular, we must refer to the *Breaking Changes* section of TypeScript's official announcement blog post.

RxJS

The RxJS versions officially supported by Angular can be read by inspecting the `dependencies` property of the `@angular/core` `package.json` file. Angular versions 9.0–10.0 officially support RxJS versions 6.5 and 6.6 while Angular versions 10.1–12.1 only have official support for RxJS version 6.6. Angular version 12.2 has opt-in support for RxJS version 7.0 and later minor versions.

Node.js

Angular CLI usually has official support for two major versions of Node.js. Unstable (odd) major version Node.js releases are not officially supported by Angular CLI. Angular CLI versions 9.0–11.2 have official support for Node.js 10.13 and 12.11 or later minor versions. Angular version 12 removes support for Node.js 10 but adds official support for Node.js 14.15 or later minor versions in addition to Node.js 12.14 or later minor versions.

In this section, we learned about the Angular Update Guide and how to manage Angular's dependencies. In the next section, we will learn about the `ng update` command and automated Angular Ivy migrations.

Performing automated Angular Ivy migrations

Angular CLI supports automated migrations for both Angular framework packages and third-party Angular libraries. In this section, we will learn how to make the most out of the `ng update` command. Finally, we will discuss important automated Angular Ivy migrations.

Making the most of the ng update command

The `ng update` command is used to update Angular-specific package dependencies, both Angular framework packages and third-party Angular libraries. The `ng update` command looks for automated migrations in the package bundle when updating to the specified package version.

To update Angular, the following command can be used:

```
ng update @angular/cli @angular/core
```

This will update all the main Angular framework packages to the latest version as well as performing their automated migrations. Angular CLI is responsible for workspace migrations while the Angular Core package is responsible for migrations to Angular's runtime packages.

In the *Learning the Angular update process* section, we recommended only updating one major version at a time. To specify, for example, Angular version 9, use the following command:

```
ng update @angular/cli^9 @angular/core^9
```

This will update the main Angular framework packages to the latest version 9 patch versions.

It is possible to perform each migration in a separate commit by specifying the `--create-commits` parameter, as shown in the following command:

```
ng update @angular/cli^9 @angular/core^9 --create-commits
```

This option is recommended as it makes it easier to inspect the changes related to each migration or use Git to cherry-pick the automated migrations we want, or even revert a migration's changes.

If we choose to revert or omit a migration through Git cherry-picking, we usually want to perform the migration manually. Alternatively, we can rerun a specific migration using the following command format:

```
ng update <package-name>[@<package-version>] --migrate-only
<migration-name>
```

We find the name of the migration in the message of the Git commits that are created when specifying the `--create-commits` parameter.

In some cases, optional migrations are available. For example, Angular version 12 introduces an optional automated migration for making the `production` build configuration the default:

```
ng update @angular/cli@^12 --migrate-only production-by-default
```

As for when to run the main `ng update` command, we follow the Angular Update Guide's instructions as described in the *Learning the Angular update process* section.

For every migration run by the `ng update` command, we see a list of files affected by the migration – if any – before `Migration completed` is displayed.

Some migrations refer to a web page describing the migration. For example, why the change is needed in addition to code snippets with examples of code before and after running the migration. This is excellent information to review the changes made by the automated migration or to perform the migration steps manually.

Reviewing automated Angular Ivy migrations

Let's review some of the most important automated Angular Ivy migrations to understand their importance.

Angular workspace version 9 migration

Named `workspace-version-9`, this migration modifies build configurations so that the `aot` option is set to `true`, even in the default development build configuration. In fact, if we generate a new Angular Ivy workspace or application using Angular CLI 12, there is no `aot` option specified because its value is `true` by default.

This migration also changes the `include` property of `tsconfig.app.json` files to match the `"src/**/*.d.ts"` pattern.

Lazy loading syntax migration

This migration, named `lazy-loading-syntax`, changes string-based lazy loaded route paths to use dynamic `import` statements instead. For example, look at the following route configuration:

```
{
  path: 'dashboard',
  loadChildren: './dashboard.module#DashboardModule',
},
```

It is changed to the following by the migration:

```
{
  path: 'dashboard',
  loadChildren: () => import('./dashboard.module')
    .then(m => m.DashboardModule),
},
```

The string-based lazy loading route syntax is deprecated and must be avoided.

Static flag migration

Be careful with this migration named `migration-v9-dynamic-queries`. In Angular version 8, the required `static` option is added to `ViewChild` and `ContentChild` queries. In Angular version 9, the `static` option is made optional, defaulting to `false`.

Consider the following Angular version 9 component:

```
import { Component, ElementRef, ViewChild } from '@angular/
core';

@Component({
  selector: 'app-hello',
  template: '
    <h1 #greeting>
      Hello, World!
    </h1>
    <div #error *ngIf="hasError">
      An error occurred
    </div>
  ',
})
export class HelloComponent {
  @ViewChild('error')
  errorElement?: ElementRef<HTMLElement>;
  @ViewChild('greeting', { static: true })
  greetingElement?: ElementRef<HTMLElement>;
}
```

In Angular View Engine version 7, before the static `option` was available, its view query properties would start out like so:

```
@ViewChild('error')
errorElement?: ElementRef<HTMLElement>;
@ViewChild('greeting')
greetingElement?: ElementRef<HTMLElement>;
```

After migrating to Angular View Engine version 8, we have the following view query properties because the `static` option is required:

```
@ViewChild('error', { static: false })
errorElement?: ElementRef<HTMLElement>;
@ViewChild('greeting', { static: true })
greetingElement?: ElementRef<HTMLElement>;
```

As you might be able to see, the Angular version 8 *static query migration* is good at guessing the best option for view query and content query properties. Queries for items nested in embedded views such as that created by a structural directive are converted to dynamic queries, that is, `{ static: false }`.

When we migrate to Angular Ivy version 9, the `static` option is optional but defaults to `false`, so we have the following view query properties:

```
@ViewChild('error')
errorElement?: ElementRef<HTMLElement>;
@ViewChild('greeting', { static: true })
greetingElement?: ElementRef<HTMLElement>;
```

Dynamic queries automatically have the `static` option removed by the Angular version 9 *static flag migration*.

> **Important Note**
>
> Query lists are not affected by the historic changes covered by this section because query lists are always dynamic.

Before migrating to Angular Ivy version 9, make sure to review your content and view queries. Refer to the *Static query migration guide* and *Dynamic queries flag migration* guides, which are still available in the Angular documentation at `https://angular.io/guide/static-query-migration` and `https://angular.io/guide/migration-dynamic-flag`, respectively, as of Angular version 12.2.

async to waitForAsync migration

This migration, named `migration-v11-wait-for-async`, renames the `async` testing callback wrapper to `waitForAsync` to avoid confusion with `async-await`. The new name better explains what happens when we wrap a test case callback in this testing function.

`waitForAsync` waits for all microtasks and macrotasks to finish before completing the wrapped test case. This is somewhat like injecting Jasmine and Jest's done callback parameter and calling it after the final asynchronous side effect in a test case.

Missing @Injectable and incomplete provider definition migration

This automated Angular version 9 migration named `migration-v9-missing-injectable` makes the following types of code changes:

- An `@Injectable` decorator is added to classes that are registered using class-based module providers.
- Incomplete Angular View Engine module providers are turned into value providers for the `undefined` value.

A class-based module provider can have one of the following formats:

```
@NgModule({
  providers: [
    DashboardService,
    {
      provide: weatherServiceToken,
      useClass: HttpWeatherService,
    },
  ],
})
export class DashboardServiceModule { }
```

If `DashboardService` or `HttpWeatherService` do not have `Injectable` decorators applied, this migration adds an `Injectable` decorator to their class definitions.

A module provider using the following format is evaluated differently by Angular View Engine and Angular Ivy:

```
@NgModule({
  providers: [
    { provide: MusicPlayerService },
  ],
})
export class MusicServiceModule { }
```

Angular View Engine evaluates the provider as the following value provider:

```
{ provide: MusicPlayerService, useValue: undefined }
```

Angular Ivy evaluates the provider as the following class provider:

```
{ provide: MusicPlayerService, useClass: MusicPlayerService }
```

Note that the preceding class provider is equivalent to the following class provider shorthand:

```
@NgModule({
  providers: [
    MusicPlayerService,
  ],
})
export class MusicServiceModule { }
```

Because of that difference between provider evaluation, this migration changes incomplete Angular View Engine providers to value providers specifying the `undefined` value.

Review all providers with the `useValue: undefined` part after this migration is run. This is most likely not the intent of our application.

Optional migration to update Angular CLI workspace configurations to production mode by default

Angular CLI version 12 generates project build configurations with `production` being the default configuration. The result of this is that we do not have to specify the `--configuration=production` parameter to the `ng build` command.

However, existing projects are not automatically migrated to use the production configuration by default. Use the optional migration named `production-by-default` to migrate existing projects to this new default setting. This is done primarily using the `defaultConfiguration` setting introduced by Angular version 12.

These are some of the most noteworthy, automated migrations to be aware of when updating from Angular View Engine to Angular Ivy. In the next section, we will discuss optional manual migrations to make sure our Angular Ivy application is in its best possible shape.

Performing manual Angular Ivy migrations

In this section, we will walk through optional migrations that put our application on track for future Angular versions. We will discuss fine-tuning initial navigation, optimizing change detection with `NgZone`, and improving the type safety of our unit tests.

Managing initial navigation

The following legacy values for the `initialNavigation` option for `RouterModule.forRoot` are removed by Angular Ivy version 11:

- `true`
- `false`
- `'legacy_enabled'`
- `'legacy_disabled'`

Angular Ivy version 11 also deprecates the `'enabled'` value but introduces the following new values:

- `'enabledBlocking'`
- `'enabledNonBlocking'` (default)

`'enabledBlocking'` is equivalent to `'enabled'` and is recommended for server-side rendering using Angular Universal. This value starts the initial navigation process before Angular creates an instance of the root component of our application but blocks bootstrapping of the root component until the initial navigation completes.

The default `'enabledNonBlocking'` value starts the initial navigation after Angular has created an instance of the root component of our application but allows the root component to be bootstrapped before the initial navigation completes. This behavior is like the `true` value, which has now been removed.

`'disabled'` is the third available, non-deprecated value. It disables the initial navigation process and defers to our application code to perform it by using the `Location` and `Router` services. This value should only be used for advanced use cases.

Optimizing change detection by configuring NgZone

When we call the `PlatformRef#bootstrapModule` method—usually in our application's main file—we can specify both compiler and bootstrap options. Bootstrap options are not listed in the Angular documentation as of Angular version 12.2. However, inline documentation is available.

Other than the traditional `ngZone` option, which allows us to disable `NgZone` entirely, the following two options are added by Angular Ivy:

- `ngZoneEventCoalescing`
- `ngZoneRunCoalescing`

They both accept a Boolean value that defaults to `false`. Both options optimize change detection for specific use cases by turning multiple change detection cycle requests in the same VM turn into a single operation, scheduled using an animation frame to synchronize change detection with the current frame rate.

Event coalescing (`ngZoneEventCoalsecing`) refers to native DOM event bubbling. For example, if multiple click event handlers are triggered by a single user click, change detection is only triggered once.

`ngZoneRunCoalescing` manages the `NgZone#run` method being called multiple times in the same VM turn.

It is a good default to enable both these options because they increase performance. However, they might change our application's behavior in certain edge cases, causing for example the *NG0100* error, `ExpressionChangedAfterItHasBeenCheckedError`, to be thrown in Angular development mode.

Because of this, take special care when enabling these bootstrap settings for our application.

Improving unit test type safety with TestBed.inject

Angular Ivy introduces the static `TestBed.inject` method, which is a strongly typed method that replaces the weakly typed static `TestBed.get` method.

The `TestBed.get` method returns a value of type `any`. In the following example, we see how this forces us to specify a type of annotation to the variable we store the returned dependency in:

```
it('displays dashboard tiles', () => {
  const dashboardService: DashboardService =
    TestBed.get(DashboardService);
  // (...)
});
```

When migrating to `TestBed.inject`, we can often omit the type annotation, as shown in the following equivalent code snippet:

```
it('displays dashboard tiles', () => {
  const dashboardService =
   TestBed.inject(DashboardService);
  // (...)
});
```

If the provided type is different from the provider token, we now must cast the returned dependency to `unknown` before casting it to the registered type, as shown in the following example:

```
it('displays dashboard tiles', () => {
  TestBed.configureTestingModule({
    providers: [
      {
        provide: DashboardService,
        useClass: DashboardServiceStub
      },
    ],
  });
```

```
  const dashboardServiceStub =
    TestBed.inject(DashboardService)
      as unknown as DashboardServiceStub;
  // (...)
});
```

It is worth noting that TestBed.inject is also stricter than TestBed.get in that it only accepts a provider token arguments of type Type<T> | AbstractType<T> | InjectionToken<T>, that is, a concrete class, an abstract class, or a dependency injection token.

This is different from TestBed.get, which supports a provider token of type any, for example a string, a number, or a symbol.

Avoid provider tokens that are not supported by TestBed.inject as they have been deprecated since Angular version 4, like the weakly typed Injector#get signature used for resolving dependencies at runtime.

Summary

In this chapter, we discussed the Angular update process, including the *Angular Update Guide*, the ng update command, and managing Angular's third-party dependencies.

We learned how to review certain important automated Angular Ivy migrations by going through simple code examples.

Finally, we considered several optional migrations, both automated and manual Angular Ivy migrations. We learned how to fine-tune the Angular router's initial navigation based on our application platform.

After that, we discussed two undocumented configuration settings for NgZone that optimize change detection by coalescing multiple requested change detection cycles into one for certain native events and use cases.

The final manual migration we discussed improves type safety in our unit tests by using the strongly typed static TestBed.inject method instead of the deprecated static TestBed.get method.

In the next chapter, we will explore the impact and limitations of the Angular Ahead-of-Time compiler, which is the default for applications in Angular Ivy.

12
Embracing Ahead-of-Time Compilation

Angular Ivy is the most recent generation of the Angular framework. It features a new compiler and a new runtime, both of which maintain compatibility with most of the APIs used by the previous generation of Angular's compiler and runtime known as Angular View Engine.

In this chapter, we will learn about how Angular Ivy makes **ahead-of-time** the default Angular compiler across all phases of development and what impact it has on our developer workflow.

Following that, we will explore metadata errors that might occur when using the ahead-of-time Angular compiler, with accompanying techniques to fix the errors.

Finally, we will demonstrate two techniques for resolving asynchronous dependencies before bootstrapping our Angular Ivy application.

We will cover the following topics in this chapter:

- Using the ahead-of-time compiler for all development phases
- Dealing with the ahead-of-time compiler's limitations
- Initializing asynchronous dependencies

After reading this chapter, you will be aware of how Angular Ivy made it feasible for the ahead-of-time Angular compiler to become the default compiler for all phases of our development workflow. You will understand how the ahead-of-time compiler affects the different phases of our development workflow.

This chapter introduces edge cases that are incompatible with the ahead-of-time Angular compiler but also teaches you how to deal with them or get around them.

You will learn two techniques for resolving asynchronous dependencies before bootstrapping your Angular application and the trade-offs they bring.

Technical requirements

The technicalities discussed in this chapter apply to applications at or higher than the following versions:

- Angular Ivy version 12.2
- TypeScript version 4.2

More metadata errors might occur in earlier versions of Angular and TypeScript.

You can find complete code examples for the feature flags and the feature flag initializer in this book's companion GitHub repository at `https://github.com/PacktPublishing/Accelerating-Angular-Development-with-Ivy/tree/main/projects/chapter12`.

Using the ahead-of-time compiler for all development phases

In previous generations of Angular, the ahead-of-time compiler was significantly slower than the just-in-time Angular compiler. Because of this and other factors, just-in-time was the default compiler in all or several phases of development depending on the Angular version. This in turn led to issues because errors were only discovered when doing a production build or—even worse—in the production environment at runtime.

Angular Ivy uses its **ahead-of-time compiler** by default in all development phases, including while running the development server, when running tests, for server-side rendering, and in the browser, instead of bundling and running the **just-in-time compiler** at runtime.

This section discusses how the ahead-of-time Angular compiler affects these phases of our development workflow.

Ahead-of-time compilation for builds

Besides improved compilation speed in Angular Ivy, another key to using the ahead-of-time compiler by default is that Angular Ivy decreases our application's bundle size under certain conditions. In general, both small and large applications see a decrease in overall size when comparing View Engine to Ivy compilation, while medium-sized applications might not see a significant change.

More specifically, the bundle sizes of small and simple applications decrease when using Angular Ivy. For complex applications, the main bundle size increases while lazy-loaded chunks become smaller when compared to Angular View Engine. This is a bit of a trade-off, considering that a bigger main bundle size increases several performance timing measurements.

The secret to unlocking the opportunity for smaller bundles is the transition from View Engine's data structures interpreted at runtime to the so-called **Ivy Instruction Set**, which reuses runtime commands—or instructions—rather than having a unique data structure for each part of your application as was the case with View Engine.

View Engine's compiled data structures have the downside that there is an inflection point at which the compiled data structures become larger than the source code of the compiler and our application.

In comparison, the Ivy Instruction Set is tree-shakable, meaning that only the instructions used by our application are included in a production bundle. For example, internationalization instructions are removed from a production bundle if our application is not multilingual. Similarly, animation instructions are excluded from a production bundle if our application does not use animations.

The Ivy Instruction Set is used by a significantly faster runtime when compared to View Engine because the precompiled instructions are immediately executable in contrast to the View Engine data structures, which must be interpreted to instructions at runtime before being executed.

Ahead-of-time compilation for component templates

When using Angular Ivy's ahead-of-time compiler, it is recommended to enable strict template type checking as mentioned in *Chapter 2, Boosting Developer Productivity Through Tooling, Configuration, and Convenience.*

Strict template type checking catches most type errors in Angular metadata, component models, and component templates. They will appear either when building our application or inline in our code editors when using the Angular Language Service.

Without strict template type checking, these errors might appear as frustrating bugs at runtime.

Ahead-of-time compilation for unit tests

Angular Ivy decreases both build time and rebuild time. This is a time-saving improvement both for the development server and for unit tests. In addition to faster compilation speed, Angular Ivy introduces a compile cache so that compiled Angular modules, components, directives, pipes, and services are cached between test cases.

In Angular View Engine, ahead-of-time compilation is not supported for unit tests. Angular Ivy introduces ahead-of-time compilation support for unit tests while still allowing the dynamic creation of Angular modules, components, directives, and pipes for the purpose of testing. Dynamic creation during unit tests still uses the just-in-time Angular compiler.

Ahead-of-time compilation for the runtime

When using the ahead-of-time Angular compiler, our Angular application is loaded faster because the just-in-time Angular compiler is not bundled with our application. Our application is bootstrapped faster because the compiler is run at build time rather than at runtime.

The Angular Ivy runtime is faster than the Angular View Engine runtime thanks to the Ivy Instruction Set. The View Engine runtime has to interpret the view compiler data structures before initializing or updating the DOM managed by an Angular component template. By comparison, Ivy instructions are executed immediately.

In this section, we discussed the impact the ahead-of-time Angular compiler has on the different phases of our development workflow. In the following section, we will shine a light on the ahead-of-time compiler's limitations and explore how we address them through simple code examples.

Dealing with the ahead-of-time compiler's limitations

The upside of using Angular Ivy's ahead-of-time compiler is faster runtime speed and a smaller bundle because of not having to ship a compiler to the runtime bundle or compiler before rendering the application.

When using the ahead-of-time compiler, there is a trade-off to be aware of. Declarables—that is, directives, components, and pipes—cannot rely on runtime information because they must be compiled ahead of time, that is, at build time rather than at runtime.

This sets a limitation for dynamically creating declarables at runtime, for example, based on server-side configuration or a static configuration file. Unless, of course, we bundle the Angular compiler with our application and use it at runtime, but then what would be the point?

The good news is that injected dependencies—that is, class-based services, provided values, or functions—can be resolved at runtime. Keep in mind that only synchronously resolved values can be provided directly. If we need an asynchronous process to resolve a value, we must wrap it in a class-based service, a function, a promise, or an observable. This is discussed and solved in the final sections of this chapter.

In this section, we will briefly discuss metadata errors when using the ahead-of-time Angular compiler. We will not discuss metadata errors that are solved by using strict TypeScript compilation or strict template type checking, as was discussed in *Chapter 2, Boosting Developer Productivity Through Tooling, Configuration, and Convenience.*

Providing values using functions

Passing the result of a function call to a value provider is not supported. Instead, we use factory functions and declare them as providers known as factory providers. For example, say we have the following ahead-of-time-incompatible **value provider**:

```
{ provide: timeZoneToken, useValue: guessTimeZone() }
```

We can replace it with the following ahead-of-time-compatible **factory provider**:

```
{ provide: timeZoneToken, useFactory: guessTimeZone }
```

This configures Angular to run the factory function at an appropriate time of our application life cycle to resolve the time zone dependency represented by `timeZoneToken` while maintaining compatibility with the ahead-of-time Angular compiler.

Declaring metadata using functions

A known ahead-of-time compilation edge case is using functions or static methods to determine declared metadata such as the `imports` or `declarations` of an Angular module.

In the following use case, we attempt to bootstrap a fake root component in Angular development mode:

```
import { NgModule, Type } from '@angular/core';
import { BrowserModule } from '@angular/platform-browser';

import { environment } from '../environments/environment';
import { AppFakeComponent } from './app-fake.component';
import { AppComponent } from './app.component';

function determineAppComponent(isDevelopment: boolean):
  Type<any>[] {
  if (isDevelopment) {
    return [AppFakeComponent];
  } else {
    return [AppComponent];
  }
}

const developmentEnvironment = environment.production ===
false;

@NgModule({
  bootstrap: determineAppComponent(developmentEnvironment),
  declarations: determineAppComponent(
   developmentEnvironment),
  imports: [BrowserModule],
})
export class AppModule {}
```

When we try to compile this code ahead of time, we hit a limitation on Angular metadata. Functions used for metadata declaration must only contain a single `return` statement, nothing else.

To comply with metadata limitations, we refactor the `determineAppComponent` function to the following implementation:

```
function determineAppComponent(isDevelopment: boolean):
  Type<any>[] {
```

```
    return isDevelopment ? [AppFakeComponent] :
      [AppComponent];
}
```

The function now contains only a single expression, a `return` statement that evaluates a ternary expression. This is compliant with the ahead-of-time Angular compiler.

Using tagged template literals in component templates

Unfortunately, ahead-of-time Angular does not support tagged template literals. For example, the following component results in a compilation error:

```
import { Component } from '@angular/core';

const subject = 'World';
const greeting = String.raw`Hello, ${subject}!`;

@Component({
  selector: 'app-root',
  template: '<h1>' + greeting + '</h1>',
})
export class AppComponent {}
```

Instead, we can use a regular function to create a compile-time dynamic part of our template, as seen in the following implementation:

```
import { Component } from '@angular/core';

const subject = 'World';
const greeting = `Hello, ${subject}!`;

@Component({
  selector: 'app-root',
  template: '<h1>' + greeting + '</h1>',
})
export class AppComponent {}
```

We must stay clear of tagged template literals for component template metadata to comply with the ahead-of-time Angular compiler's limitations on Angular metadata. However, we can use tagged template literals in UI properties that we use in template bindings, as seen in the following refactored implementation:

```
import { Component } from '@angular/core';

const subject = 'World';

@Component({
  selector: 'app-root',
  template: '<h1>{{ greeting }}</h1>',
})
export class AppComponent {
  greeting = String.raw`Hello, ${subject}!`;
}
```

Choose either of these two techniques to deal with or support tagged template literals for Angular component templates, respectively.

Initializing metadata variables

Metadata must be immediately available for the ahead-of-time compiler. The following example is invalid because of late initialization:

```
import { Component } from '@angular/core';

let greeting: string;
setTimeout(() => {
  greeting = '<h1>Hello, World!</h1>';
}, 0);

@Component({
  selector: 'app-hello',
  template: greeting,
})
export class HelloComponent {}
```

The `greeting` variable has not been initialized when the component metadata is converted to annotations by the ahead-of-time Angular compiler.

More surprisingly, the following example is also invalid:

```
import { Component } from '@angular/core';

let greeting: string;
greeting = '<h1>Hello, World!</h1>';

@Component({
  selector: 'app-hello',
  template: greeting,
})
export class HelloComponent {}
```

While not a common use case, keep this limitation in mind because it is quite surprising.

Let's first see the fixed implementation:

```
import { Component } from '@angular/core';

let greeting = '<h1>Hello, World!</h1>';

@Component({
  selector: 'app-hello',
  template: greeting,
})
export class HelloComponent {}
```

The `greeting` variable is now defined and initialized at the same time so the component template works as expected.

Let's now change the value of the `greeting` variable just after initializing it:

```
import { Component } from '@angular/core';

let greeting = '<h1>Hello, World!</h1>';
greeting = '<h1>Hello, JIT compiler!</h1>';

```

```
@Component({
  selector: 'app-hello',
  template: greeting,
})
export class HelloComponent {}
```

When using the ahead-of-time Angular compiler, the template is `<h1>Hello, World!</h1>`. If we change to the just-in-time Angular compiler, the template is `<h1>Hello, JIT compiler!</h1>`.

Variables used for declarable metadata must be defined and initialized at the same time when using the ahead-of-time Angular compiler.

In this section, we explored the edge cases of ahead-of-time compiler compatibility and learned how to address them. In the next section, we will learn how to initialize asynchronous dependencies before bootstrapping our application.

Initializing asynchronous dependencies

Referring to asynchronous values is toxic because every value computed from the referred value must be asynchronous as well. A couple of techniques to get around this are available, but they both come at the cost of delaying application bootstrapping until the value has been resolved. These techniques are demonstrated in this section.

Providing an asynchronous dependency with a static platform provider

To convert an asynchronous dependency resolver to a static dependency, we can delay bootstrapping our application to provide the static provider at the platform level, making it available as a static dependency in our application.

For example, say we have a JSON file containing an object with Boolean values. We create it in the `src/app/assets/features.json` file of our application project. This file contains our **feature flags**, which are loaded at runtime. The settings in this file can be changed after compiling our application.

In the `src/load-feature-flags.ts` file of our application project, we add the following function:

```
export function loadFeatureFlags():
  Promise<{ [feature: string]: boolean }> {
  return fetch('/assets/features.json')
```

```
      .then((response) => response.json());
}
```

Before calling this function in the main file of our application, we create a dependency injection token to represent the feature flags.

The following code block shows the `src/app/feature-flags.token.ts` file of our application project:

```
import { InjectionToken } from '@angular/core';

export const featureFlagsToken =
  new InjectionToken<Record<string, boolean>>(
    'Feature flags'
  );
```

Finally, we modify our main file so that it contains something like the following:

```
import { enableProdMode } from '@angular/core';
import { platformBrowserDynamic } from '@angular/platform-
browser-dynamic';

import { AppModule } from './app/app.module';
import { featureFlagsToken } from './app/feature-flags.token';
import { environment } from './environments/environment';
import { loadFeatureFlags } from './load-feature-flags';

if (environment.production) {
  enableProdMode();
}

loadFeatureFlags()
  .then((featureFlags) =>
    platformBrowserDynamic([
      { provide: featureFlagsToken, useValue: featureFlags
      },
    ]).bootstrapModule(AppModule)
  )
  .catch((err) => console.error(err));
```

Notice how we are loading the feature flags, then passing them as the value for the platform provider of the feature flags token before being able to bootstrap our Angular application module.

Now, we can inject the feature flags in any Angular-specific class, such as a component, as seen in the following example:

```
import { Component, Inject } from '@angular/core';

import { featureFlagsToken } from './feature-flags.token';

@Component({
  selector: 'app-root',
  template: `
    <div *ngFor="let feature of features | keyvalue">
      <mat-slide-toggle [checked]="feature.value">
        {{ feature.key }}
      </mat-slide-toggle>
    </div>
  `,
})
export class AppComponent {
  constructor(
    @Inject(featureFlagsToken)
    public features: { [feature: string]: boolean }
  ) {}
}
```

Feature flags are a good use case for this technique. Other configurations are also well suited for this approach. Additionally, if multiple application initializers need a shared dependency, this technique is the best approach.

In the next section, we will walk through an alternative technique and discuss the differences.

Resolving an asynchronous dependency with an application initializer

Another technique for dealing with a dependency that is asynchronously resolved is an application initializer.

An application initializer is resolved before the root application component is bootstrapped. This is ideal for setting up the initial root-level state that is not needed for other application initializers.

We will consider an alternative approach for dealing with feature flags. This time, we are using a feature flag service that is configured using an application initializer.

The feature flag service has the following implementation:

```
import { Injectable } from '@angular/core';

@Injectable({
  providedIn: 'root',
})
export class FeatureFlagService {
  #featureFlags = new Map<string, boolean>();

  configureFeatures(featureFlags: { [feature: string]:
   boolean }): void {
    Object.entries(featureFlags).forEach(([feature, state])
      =>
      this.#featureFlags.set(feature, state)
    );
  }

  isEnabled(feature: string): boolean {
    return this.#featureFlags.get(feature) ?? false;
  }
}
```

The feature flag initializer loads the feature flags using `HttpClient` before calling `FeatureFlagService#configureFeatures`. This is seen in the following code listing:

```
import { HttpClient } from '@angular/common/http';
import { APP_INITIALIZER, FactoryProvider } from '@angular/
core';
import { Observable } from 'rxjs';
import { mapTo, tap } from 'rxjs/operators';

import { FeatureFlagService } from './feature-flag.service';

function configureFeatureFlags(
  featureFlagService: FeatureFlagService,
  http: HttpClient
): () => Observable<void> {
  return () =>
    http.get<{ [feature: string]: boolean
     }>('/assets/features.json').pipe(
      tap((features) =>
       featureFlagService.configureFeatures(features)),
      mapTo(undefined)
    );
}

export const featureFlagInitializer: FactoryProvider = {
  deps: [FeatureFlagService, HttpClient],
  multi: true,
  provide: APP_INITIALIZER,
  useFactory: configureFeatureFlags,
};
```

Finally, we register the feature flag initializer in our root module by adding it to the `providers` array, as seen in the following code snippet:

```
import { HttpClientModule } from '@angular/common/http';
import { NgModule } from '@angular/core';
import { MatSlideToggleModule } from '@angular/material/slide-
```

```
toggle';
import { BrowserModule } from '@angular/platform-browser';

import { AppComponent } from './app.component';
import { featureFlagInitializer } from './feature-flag.
initializer';

@NgModule({
  bootstrap: [AppComponent],
  declarations: [AppComponent],
  imports: [BrowserModule, HttpClientModule,
   MatSlideToggleModule],
  providers: [featureFlagInitializer],
})
export class AppModule {}
```

After setting all of this up, any Angular-specific class can inject an instance of the `FeatureFlagService` class and use its `isEnabled` method to check the state of a feature flag, as seen in the following code listing:

```
import { Component } from '@angular/core';

import { FeatureFlagService } from './feature-flag.service';

@Component({
  selector: 'app-root',
  template: `
    <div>
      <mat-slide-toggle
        [checked]="featureFlagService.isEnabled(
         'middleOutCompression')"
      >
        Middle-out compression
      </mat-slide-toggle>
    </div>

    <div>
```

```
        <mat-slide-toggle
          [checked]="featureFlagService.isEnabled(
           'decentralized')"
        >
          Decentralized application
        </mat-slide-toggle>
      </div>
    `,
})
export class AppComponent {
  constructor(public featureFlagService:
    FeatureFlagService) {}
}
```

The benefit of using an application initializer is that multiple of them can be run in parallel, which speeds up the overall application bootstrap time compared to delaying the entire bootstrap process until a response has finished, as was the case in the preceding section.

The trade-off is that we must wrap the feature flags in a service-based class with methods for writing and reading the feature flag configuration, whereas with the first technique we explored, the feature flags were available as a static dependency, a dead simple object. Choose whichever technique fits your use case best.

Summary

In this chapter, we learned how the enhanced Angular Ivy compiler and runtime make the ahead-of-time Angular compiler a good choice for all phases of development. The tree-shakable, reusable Ivy Instruction Set leaves a smaller bundle for a range of applications.

We discussed how ahead-of-time compilation affects our application builds, component templates, unit tests, and the browser at runtime.

Next, we explored solutions for metadata errors that occur when using the ahead-of-time Angular compiler. Metadata errors that are detected by strict TypeScript and Angular compilation settings were not discussed. Read about strict template type checking in *Chapter 2, Boosting Developer Productivity Through Tooling, Configuration, and Convenience.*

In the final sections, we learned how to resolve and initialize asynchronous dependencies before bootstrapping our application using two techniques:

- Providing an asynchronous dependency with a static platform provider
- Resolving an asynchronous dependency with an application initializer

These techniques are great for feature flags and other configurations, but they each have trade-offs that you are now able to recognize.

That is the end of this book. We hope that you enjoyed learning about some of the most interesting stable features introduced by Angular Ivy and its accompanying versions of TypeScript. Angular is an ever-evolving framework with several feature releases every year.

Keep learning!

Other Books You May Enjoy

If you enjoyed this book, you may be interested in these other books by Packt:

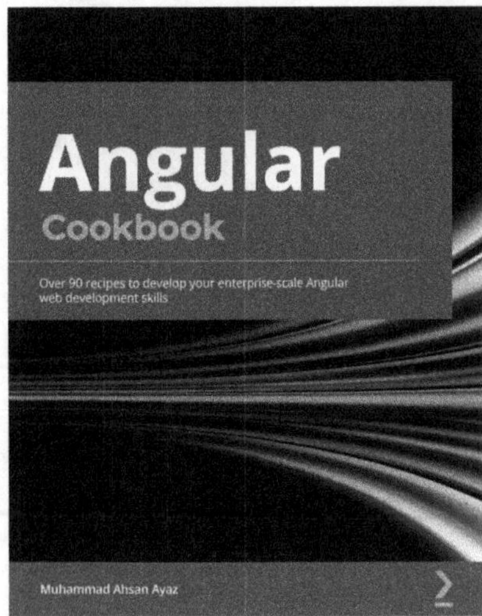

Angular Cookbook

Muhammad Ahsan Ayaz

ISBN: 978-1-83898-943-9

- Gain a better understanding of how components, services, and directives work in Angular
- Understand how to create Progressive Web Apps using Angular from scratch
- Build rich animations and add them to your Angular apps
- Manage your app's data reactivity using RxJS
- Implement state management for your Angular apps with NgRx

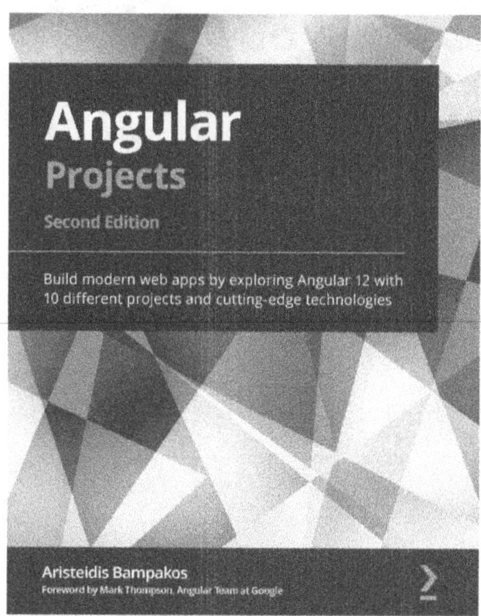

Angular Projects - Second Edition

Aristeidis Bampakos

ISBN: 978-1-80020-526-0

- Set up Angular applications using Angular CLI and Nx Console
- Create a personal blog with Jamstack and SPA techniques
- Build desktop applications with Angular and Electron
- Enhance user experience (UX) in offline mode with PWA techniques
- Make web pages SEO-friendly with server-side rendering
- Create a monorepo application using Nx tools and NgRx for state management
- Focus on mobile application development using Ionic
- Develop custom schematics by extending Angular CLI

Packt is searching for authors like you

If you're interested in becoming an author for Packt, please visit `authors.packtpub.com` and apply today. We have worked with thousands of developers and tech professionals, just like you, to help them share their insight with the global tech community. You can make a general application, apply for a specific hot topic that we are recruiting an author for, or submit your own idea.

Hi!

We are Lars Gyrup Brink Nielsen and Jacob Andresen, authors of *Accelerating Angular Development with Ivy*. We really hope you enjoyed reading this book and found it useful for increasing your productivity and efficiency in Angular.

It would really help us (and other potential readers!) if you could leave a review on Amazon sharing your thoughts on *Accelerating Angular Development with Ivy*.

Go to the link below or scan the QR code to leave your review:

`https://packt.link/r/180020521X`

Your review will help us to understand what's worked well in this book, and what could be improved upon for future editions, so it really is appreciated.

Best Wishes,

Jacob Andresen

Index

A

B

C

D

E

F

G

H

I

L

M

N

O

P

R

S

T

U

Z

www.ingramcontent.com/pod-product-compliance
Lightning Source LLC
Chambersburg PA
CBHW082117070326
40690CB00049B/3605

* 9 7 8 1 8 0 0 2 0 5 2 1 5 *